PRAISE

BECOM

PSYCHIC

"The fascinating Dr. Jeff Tarrant walks you through the many doors he has traveled to find the ultimate connections in *Becoming Psychic*. His research, personal experiences, and professional understanding of the brain demonstrate what we are capable of when we are open-minded. He shares tips and tools to help one gain understanding into the complex, multifaceted discoveries of the brain that demonstrate the potential to be life-altering."

—**Janet Mayer,** psychic medium, author of *Spirits . . . They Are Present*

"Good science often begins with subjective experiences and inquisitiveness of the scientist. Dr. Tarrant has spent much of his professional career exploring nonphysical phenomena that materialist science has been unable to explain. By focusing on the human brain and examining changes that occur when one is exhibiting consciousness that extends beyond the body, his work has shown us that there are ways we can all achieve what many believe to be impossible. Unlocking the filters of the brain can open us up to access unseen worlds and information—in essence, putting the mystical realm within our grasp. *Becoming Psychic* provides a roadmap for those who suspect that there is more than meets their eyes."

—**Bob Ginsberg,** founder, Forever Family Foundation,
and author of *The Medium Explosion*

"In *Becoming Psychic*, clinical psychologist Jeffrey Tarrant, PhD, shares his deeply personal exploration into the complex and fascinating realm of psychic phenomena. As a scientifically minded researcher, Tarrant was initially skeptical of the existence of psychic abilities, but through his own experiences with psychic mediums and other intuitive practices, his perspective began to shift. In this compelling book, Tarrant shares his journey from skeptic to believer, exploring the scientific research behind psychic phenomena and offering practical tools, grounded in his own experience, for accessing and harnessing one's own intuitive abilities. Through this compelling personal story and expert insights, Tarrant's readers will gain a deeper understanding of the mind-body-spirit connection and the potential for personal growth and transformation."

—**Marilyn Schlitz, PhD, MBA,** professor, Sofia University, author of *Death Makes Life Possible*

"*Becoming Psychic* is the engaging personal journey of Jeff Tarrant's encounters with extraordinary people, accompanied by his scientific studies of their brain activity. A fascinating read!"

—**Dean Radin, PhD,** chief scientist, Institute of Noetic Sciences, and author of *Real Magic* and other books

"Jeff's stories, wisdom, and clinical experiences invite us all to explore our greatest potential as human beings in a grounded and relatable way. *Becoming Psychic* opens the doors to how we can approach being with ourselves and each other in ways that were historically denied or feared. Jeff shows us how curiosity, an open mind, and a bit of courage are the medicine that shift the paradigm of how we engage with the world in which we live."

—**Jeannine Kim,** intuitive healer, astrologer, mystic, and medium, and author of *Dark Matters*

"*Becoming Psychic* by Jeff Tarrant provides a diverse and balanced look into the realms of unseen forces and psychic abilities, combining scientific-based knowledge with the hands-on experience of psychic practitioners from various fields, including psychokinesis, channeling, and healing. Jeff's willingness to delve into the unknown with an open mind is admirable and combined with his extensive knowledge in the fields of meditation and breathwork, leave the reader with a much greater understanding of the workings within the supernatural realm. If you are interested in psychic phenomenon and wish to hear both sides of the story, this book is a great addition to your library."

—**Robert Allen, AKA Trebor Seven,** psychokinesis teacher and educator

"In *Becoming Psychic,* Jeff Tarrant presents techniques and exercises based on his brain research and work with mediums and psychics and his own experiences, offering pathways for the rest of us to awaken our own psychic potential. While there is a plethora of psychic development books out there, none I've seen are solidly based in scientific research, and certainly none cover what's going on in our heads when we try different practices and techniques, and when we are and are not psychic. I highly recommend this book to anyone who wants a grounded approach to becoming psychic."

—**Loyd Auerbach,** parapsychologist and paranormal investigator, and author of *Psychic Dreaming: Dreamworking, Reincarnation, Out-of-Body Experiences & Clairvoyance* and other books

"Finally, a book that delivers groundbreaking revelations, together with irrefutable data to improve our understanding of energy and the function of the brain during the altered states of consciousness of psychics and mediums. With his open-minded approach, together with several scientific experiments, Tarrant delivers an easy-to-read masterpiece that is sure to alleviate one of the biggest fears of all time—the fear of death! This comprehensive work of art brings a unique and refreshing perspective to the most widely researched topics—the brain and consciousness! Whether you are a scientist yourself, a skeptic, or a diehard believer, this book is for everyone, especially the truth-seekers. As a psychic medium myself, I believe this book will bring us one step closer to uncovering a controversial truth: we don't die."

—**Kim Russo,** The Happy Medium, international psychic medium, TV host, and author

"Jeff Tarrant does a brilliant job at covering a wide range of psychic abilities and providing rigorous scientific data to support the phenomenon. The collected detail and results push the research of psychic phenomena to a whole new level of understanding. I highly recommend this book to anyone interested in the field of parapsychology, psychic abilities, and the science of consciousness."

—**Caroline Cory,** consciousness studies researcher and award-winning filmmaker *Superhuman: the Invisible Made Visible and A Tear in the Sky*

JEFF TARRANT, PhD, BCN

Foreword by **LAURA LYNNE JACKSON,**
bestselling author of *Signs* and *The Light Between Us*

BECOMING
PSYCHIC

Lessons from the
Minds of Mediums, Healers,
and Psychics

Health Communications, Inc.
Boca Raton, Florida

www.hcibooks.com

Library of Congress Cataloging-in-Publication Data
is available through the Library of Congress

© 2023 Jeff Tarrant, PhD, BCN

ISBN-13: 978-07573-2478-9 (Paperback)
ISBN-10: 07573-2478-9 (Paperback)
ISBN-13: 978-07573-2479-6 (ePub)
ISBN-10: 07573-2479-7 (ePub))

All rights reserved. Printed in the United States of America. No part of this publication may be reproduced, stored in a retrieval system, or transmitted in any form or by any means, electronic, mechanical, photocopying, recording, or otherwise, without the written permission of the publisher.

HCI, its logos, and marks are trademarks of Health Communications, Inc.

Publisher: Health Communications, Inc.
        301 Crawford Boulevard, Suite 200
        Boca Raton, FL 33432–3762

Cover, interior design, illustrations and formatting by Larissa Hise Henoch

CONTENTS

CHAPTER SEVEN
PSYCHOMANTEUM

CHAPTER EIGHT
PSYCHICS AND THE PSYCHEDELIC BRAIN

CHAPTER NINE
PSYCHOKINESIS

CHAPTER TEN
ENERGY HEALING

CHAPTER ELEVEN
LESSONS LEARNED

FOREWORD

EVER SINCE I WAS A CHILD, I was aware that I had a way of seeing people and energy and information that was not the norm. I would see people in colors, feel what they were feeling, and at times know information about things I couldn't logically have known: things that had happened in the person's past, would happen in the future, or names and information about loved ones close to them who had crossed to the other side.

As I grew older, the materialistic paradigm that we are raised with issued a judgment on me: I had to either be deluding myself into believing that I could do things I couldn't do—or a fraud—because there was no way this could be real.

Only it was real.

I knew I wasn't delusional or a fraud. I was raised by two teachers who taught me to be a critical thinker. And a critical thinker doesn't just accept what someone else states as truth and then close their mind. A critical thinker goes on a journey of discovery and seeks the truth.

And so I did.

My journey led me first to see a psychiatrist, to whom I confessed that at times I would hear voices speaking to me. I explained that it was an internal hearing—not an external sound or voice. And I told him that these voices would tell me things. "What kind of things do they tell you, Laura?" the psychiatrist asked, narrowing his gaze at me. But what the voices would tell me were never scary: they would be someone's grandmother telling me that they were happy about the new baby in the family's arrival who was named after them, or that the person I was talking to was about to get a wonderful new job.

"The world is full of more things than we can possibly imagine," he told me. He assured me that I was not crazy or delusional. "It sounds like the information you are getting is healing and beautiful," he said, "Carry on."

And so I did.

My path of discovery next led me to the world of science to try to bridge the gap between the information I was getting and what we know of consciousness; of energy both seen and unseen. I was interested in finding out how I was able to easily access this information while other people seemingly could not. I knew that if I could do it others must have this ability as well. Perhaps there was even a way they could learn to access it and use it as a tool to help guide them into their highest and best paths in life!

And then the Universe cued in Dr. Jeff Tarrant.

I first met Dr. Tarrant in 2013 at a Forever Family Foundation conference in San Diego, California. The Forever Family Foundation is a nonprofit dedicated to helping people in grief. It was founded by Bob and Phran Ginsberg after their daughter, Bailey, died. The Ginsbergs started having signs and communications from Bailey showing them she was still present and loving them—just in a new way. So, they founded this organization to help people whose loved ones had passed by connecting them with a medium (an individual who can connect with the consciousness of those no longer in a physical body). They were careful to ensure the integrity of the program by implementing strict protocols and doing blinded tests on mediums to make sure that they were proficient using them, thus eliminating any chance of fraud. Mediums had to pass their test in order to volunteer. I passed the test in 2005, and I have been volunteering ever since.

Dr. Tarrant was attending the Forever Family Foundation conference as a presenter when I met him. He had studied and mapped the brain-wave activity of one of the Forever Family Foundation certified mediums, Janet Mayer, and was presenting his unique findings.

One thing led to another, and before I knew it, I was hooked up to an EEG, having my brainwaves studied by Dr. Tarrant. This was the first of a number of occasions where he would gather data on what my brain was doing during normal talking mode, as well as when I went into psychic and mediumistic mode.

Some very interesting things were revealed.

For example, when I read, I get an inner screen that appears in my mind's eye. That screen is divided into two sections. The left-hand side of the screen is where I get all my psychic information; the right-hand side of the screen is where I get all my mediumistic information. To clarify, when I read psychically, I am reading a person's energy field and picking up on past, present, and future events, connections, etc. However, when I read mediumistically, I see points of light appear and I communicate with the consciousness of those who are no longer in a physical body.

What was both fascinating and validating was that the EEG revealed that when I was saying I was reading psychically (on the left-hand side of my screen), Dr. Tarrant was able to map my brain-wave activity. It showed that I was seeing something in my left field of vision—something that was registering in my brain but that no one else could see!

Conversely, when I switched over to reading mediumistically (on the right-hand side of my screen), the EEG once again revealed that I was now "seeing" something in my right field of vision—even though nothing appeared in the room. Dr. Tarrant's research was able to show that what I was saying I was seeing on my screen was reflected by my brain-wave activity.

Now, how does this pertain to you? The data shows that certain parts of our brains are used when tuning in psychically and mediumistically—and that we can learn how to train our brains and access this information!

Dr. Tarrant, with his EEG machines, has been able to look inside the brains of psychic mediums and see exactly what we are doing. He has found characteristic similarities, and he will share what he has found with you in this book! What's more, you will come away with an understanding of exactly how to train your brain waves to more fully access these parts of yourself! The results can be life-changing, both in seemingly small ways (getting parking spots easily) to much larger ones (knowing what job offer to accept, or if you should move). Becoming psychic can shift your life in positive and beautiful ways!

So, get ready! You are about to go on an amazing adventure through these pages —a journey that will give you an understanding of your own potential and ability to become psychic!

—**Laura Lynne Jackson**
New York Times bestselling author of *Signs*
and *The Light Between Us*

PROLOGUE: FROM SKEPTIC TO BELIEVER

LOOKING BACK, I can clearly trace the path of how I became a researcher examining the brain waves of mediums. It was not a direct path, to be sure, but it seems rare when life journeys are linear (particularly when you are talking about communicating with the dead).

As a young child I was captivated by any stories or movies that had anything to do with the supernatural. While I was not much of a reader as a child, when we went to the school library, I would look for books on Bigfoot and the Loch Ness Monster or I would check out science-related books that were way over my head. My grandmother, whom I would visit semi-regularly throughout childhood, read tabloid publications like the *National Enquirer*. When I visited her, I would thumb through her collection looking for articles about demonic possession, ghosts, or aliens. There was usually at least one article on one of these topics in every issue. I cut out these articles and saved them, believing I would need them in the future as part of my "research" into understanding these phenomena.

I read the book, *A Wrinkle in Time* by Madeleine L'Engle and became fixated on the idea of alternate realities and "magical powers." For weeks, I would lie in bed each night after reading it, trying to turn on the lights with my mind (I wasn't successful, in case you were wondering). I desperately wished for magical powers, particularly telekinesis.

When I was growing up, the big movies were *Star Wars, E.T., Close Encounters of the Third Kind,* and *Poltergeist,* which likely helped fuel my strong interest in UFOs and aliens. While most of my friends and neighbors in Arnold, Missouri, were playing sports or pursuing interests related to cars, I spent a great deal of time with my nerdy friend, Matthew. Although we both played on the soccer and baseball teams, that was not where our hearts were—maybe because we were also not very good. We loved science fiction. We would play games where we would gather our various instruments and pretend that we were conducting research on UFO landings. We would ride our bikes to neighboring areas and conduct our research on the various mountains of dirt, "testing" for a previous UFO landing. As I got a bit older, I began collecting comic books and playing Dungeons and Dragons, more age-appropriate mechanisms to continue exploring similar ideas and concepts.

During junior high and high school, I was much too interested in girls and fitting in to care about aliens, ghosts, and goblins. In fact, between the ages of fourteen and twenty, I took a significant hiatus from anything that did not involve partying. That changed in my sophomore year of college at the University of Missouri, when I became friends with three women, one of whom had a significant influence on me. Tracy was an English major and very much into

New Age spiritual thought, which I knew nothing about. She was constantly reading something about auras or astral projection or crystals. She did tarot readings, believed in fairies, and saw synchronicity in everyday events. For instance, one Sunday afternoon I was spending time with Tracy and the rest of our friend group, which was a typical scene. Tracy told us that she had invited a new friend, Steve, to join us. I had met him before, and I did not like him. I found him to be a bit of a meathead and was jealous of his gym body and the attention he got from the other girls in our group. Tracy knew I didn't care for him and tried to convince me that we would like each other once I got to know him better. Just before he was supposed to arrive, Steve called to cancel, saying he had a bad headache and wouldn't make it. Of course, I was elated at this turn of events, but Tracy was convinced that *I gave* Steve the headache. While this might sound somewhat delusional, this was the way Tracy thought. She saw the connection between things and was convinced that our thoughts can influence those around us. Little did I know that I would be studying this idea in a more scientific way many years later.

Tracy and I became fast friends. We just connected, and our relationship rekindled some of my childhood interests and beliefs. This time, however, things were different; it was more applied. Rather than just pretending or engaging in fantasy about concepts like telepathy and clairvoyance, I wanted to learn how to do it. I wanted to develop these skills, but I didn't know how. I tried to learn to use the tarot cards but didn't have the patience for it. I wanted to meditate but had no idea what I was doing and could not find any teachers in mid-Missouri in the late 1980s.

The one skill I did learn from Tracy and enjoyed immensely was seeing auras. It is actually fairly easy and involves unfocusing the eyes and looking at the edges of the person or object. I started doing this with my coworkers at Madison's Café, where I worked as a line cook. I would sit in the back of the auditorium in my classes and look at the aura of my professors. Most of the time, I simply saw a clear- or whitish-colored light close to the person's body, but other times I would see yellow, green, or blue, which was very exciting. I had no idea what any of it meant, but it suggested that something about this was real and beyond the scope of so-called "normal reality," whatever that is.

As interested as I was in the supernatural, part of me thought there was a logical explanation for these experiences; perhaps seeing auras is merely a trick played on the visual circuitry of the brain and nothing more. However, as my ability to read auras sharpened and felt more natural to me, I began to see a variety of colors and also started to notice an increasing number of synchronicities in my daily life. I then became more convinced that there was something going on that was real outside of our normal perceptions—and very interesting.

Unfortunately, my circle of New Age friends graduated from college, and we went separate ways. We lost touch with each other, and I shifted back to a more earthly existence. While continuing to go to school to get my master's and then a doctoral degree in counseling psychology, I got married and we started a family. The demands of school, a part-time job, and raising two young children didn't leave me with a lot of free time.

During my graduate training, I continued to be interested in the paranormal, which took the form of becoming a huge fan of the *X-Files* and *Star Trek: The Next Generation*. We regularly had friends over, and I would inevitably launch into a rather complicated and sophisticated discussion regarding the evidence for extraterrestrial life, the fact that they visit us, and that alien abductions are happening. Most of my friends were tolerant—maybe even somewhat interested—although I don't believe any of them took me very seriously, likely just thinking this was one of my quirks.

As I continued my graduate training and began to adopt a more professional identity, something changed. I became more cynical and skeptical. I wrote off a vast majority of these beliefs as wishful thinking. I looked at my own childhood games and fantasies as psychological attempts to escape, to feel powerful. My stepfather, who was in my life from ages nine until twenty, was emotionally abusive. A Vietnam War veteran, he dealt with his trauma by drinking, and he was controlling, unhappy, and unpredictable. It made sense that I had adopted beliefs in supernatural powers. They created the hope that I could be powerful and special. My frequent fantasies about being taken aboard an alien spacecraft and whisked away to a foreign planet also made sense. In my mind, I could escape my situation and start a new life.

I began reading magazines like *The Skeptical Inquirer* and *Skeptic,* which presented arguments that made sense to me; humans are simply very good at deluding and fooling themselves. Rational and logical thinking dictates that we see belief in ghosts and paranormal abilities as psychological reactions and tricks of the mind. My psychology training program, which was very research-based, emphasized this left-brained way of thinking. In fact, their approach to counseling psychology was called "the scientist-practitioner model." Essentially, it was shaping and training me into a particular way of viewing the world; that an atheistic, logical, and rational approach to the world was the only one with validity. Belief in the paranormal was associated with immaturity (at best) and psychopathology (at worst). Consciousness was seen as arising from the neural connections of the brain. When we stop breathing and the brain ceases its functions, consciousness is lost and the body decays. There is no God. There are no ghosts. Nothing is real unless science can prove it. Basically, all of this is to say that my thinking was conditioned, not overtly but subtly over time. The same thing happens in every other professional field; I've seen it happen with friends and colleagues. The training to be a medical doctor or lawyer changes you: in both the way you perceive the world and what you believe to be true. In my case, I was no longer open to experiences and possibilities that were outside the realm of what was considered "normal." My rigid, skeptical view persisted for many years, but over time my interest in the supernatural and my openness to all things weird returned (thank goodness).

I'm not sure exactly how or when my mind opened again, but it was probably connected to my exploration of spirituality. Sometime in the late 1990s, I became interested in meditation and the idea of exploring states of consciousness. I had been practicing martial arts for six or seven years at this point and loved the idea of mastering my internal world through disciplined practice. I explored yoga and *qigong* (pronounced "chi gong") and began learning about *qi* (also known as *chi*) and *prana*— the subtle energies that are within and around us. I began meditation training with a Zen monk and learned to quiet my scattered mind.

At the same time, my practice as a psychologist had shifted away from talk therapy and psychological assessments to the field of *neurofeedback*. This involves measuring a person's brain wave activity to help identify patterns that may be related to their concerns, such as depression, anxiety, or attention deficit hyperactivity disorder (ADHD). Once those patterns are identified, the client returns to the office twice a week where we attempt to train the brain toward more adaptive and flexible patterns. When effective, this process very often leads to a significant reduction of symptoms.

While doing this work, it occurred to me that I could use this technology to begin exploring changes in the brain in relation to different complementary and alternative therapies. This was a way that I might be able to explain the unexplainable. Maybe I could prove that things like meditation and directed intention had a real impact, their effects weren't just placebo or wishful thinking.

I was like a kid in a candy store, measuring every brain I could find. I looked at what was happening during various styles of meditation, when using essential oils, after movement practices such as qigong or Brain Gym, and while using technology interventions such as *audio-visual entrainment*. At some point, I realized that I could use neurofeedback and other neuromodulation technologies to encourage or coax the brain into specific meditative states. This technology-enhanced meditation process could be used to facilitate the positive impacts of meditation, making it a powerful tool to help with a variety of mental health concerns. In 2016, I founded the NeuroMeditation Institute and began teaching this approach to other practitioners. Our training program has grown quickly and now has centers in the United States, Germany, and Poland. While the NeuroMeditation approach has been primarily focused on using specific meditation strategies to reduce anxiety, depression, and post-traumatic stress disorder (PTSD), we have also begun to study and explore nonordinary states of consciousness. What happens in the brain during intense breathwork sessions, *vibroacoustic* meditations (translating music and sound into vibrations experienced through a specially designed massage table or backpack), *stroboscopic* light (light pulsations at different frequencies that influence brainwave activity), or ayahuasca retreats? How do the brain changes accompanying these practices lead to deep healing, and how can we use this information to help others? All of these investigations, curiosities, and journeys inevitably led me to an interest in exploring the brains of psychics, mediums, and energy healers. If we can use our understanding of the brain to manage mental health concerns, deepen meditative states, and increase peak performance, maybe the same approach could help us understand and develop super-ordinary states of consciousness. After measuring the brains of dozens of highly gifted individuals over a ten-year span, I am now convinced that energy healing, mediumship, ESP, telepathy, and psychokinesis are real. These abilities are not always consistent, and the results are not always mind-blowing. However, I have seen enough to convince me that we are capable of much more than most of us dare to imagine.

Of course, some of my curiosity and explorations are personal. I wanted to know if I could apply what I have learned to increase my own and others' psi abilities. Perhaps if I could uncover the brain patterns associated with psychic abilities, I could use technology to unlock or enhance these abilities.

In the next eleven chapters, I will take you through my journey, introducing you to the people I have met, the things I have witnessed, my own attempts to cultivate psi abilities, and the brain wave data I have collected. I will share the patterns I have observed and the tools, meditations, and practices that I have found to help in becoming psychic. What you do with this information is up to you, should you desire to develop, boost, or rediscover your own psychic abilities.

CHAPTER ONE

CHANNELING SHAMANS

IN 2013, my exposure to people with "extraordinary abilities" was virtually nonexistent—but that was all about to change.

At the time, I was working at the University of Missouri as a health psychologist in the student health center. In this position, I was largely responsible for developing a campus-wide, stress-management, biofeedback program. After about a year of piloting this program, we received a nice-sized grant that allowed us to purchase biofeedback software and sensors for all of the computer labs on campus. The grant also allowed me to hire two undergraduate students to assist in the program; enter Kelcie and Matt. I met them about a year before when they attended one of our early biofeedback courses. They were motivated and quick learners, so I was excited to make them part of the team. During their tenure with our program, I saw them several times a week and got to know them quite well. At the end of their appointment, they were both graduating and getting ready to move into the next phase of their lives.

At a farewell/thank you dinner, Matt announced that he wanted to tell us about his mother. He seemed nervous and cautious, and I had absolutely no idea what was coming next. *Why was he acting so weird?* I felt like he was getting ready to tell me some bad news and I was bracing myself. Matt proceeded to tell us a long and complicated story about his mother, Janet Mayer. Apparently, she began spontaneously speaking South American tribal languages several years before this conversation, after participating in a *holotropic breathwork* session. What a relief! No bad news after all. Wait, what!? Spontaneously speaking South American tribal languages after participating in holotropic breathwork? I can see why he would be cautious about sharing this news. This sounded a bit crazy. Apparently, he had waited an entire year to tell me about this because he wanted to make sure I would be open to the story. I guess I passed the test.

Holotropic Breathwork

If you are not familiar with it, holotropic breathwork or HB is a deep and rapid breathing technique that is used in combination with evocative music to induce an altered state of consciousness. Typically, the process is done with a partner in a group setting. One person acts as a "sitter" and is there for anything the breather might need during the session: help getting to the bathroom, Kleenex, a blanket. The breather inhales deeply through the nose and lets all the air out through the mouth, repeating this with no pauses and at a somewhat quickened pace. You keep this breathing going for an hour or more. Typically, after twenty to thirty minutes, consciousness shifts into a full-blown psychedelic state. This state is enhanced by the setting created by the facilitators and the sounds and energy of the other people in the room. Having participated in a few of these myself, I have seen people writhing on the floor, crying, yelling, curled in a ball, and laughing hysterically (not all at the same time, of course). The experience often results in a release of emotions tied to previous hurts and traumas, or realizations about oneself that can be used for personal growth. For most people, it is a powerful and intense process.

Matt explained that his mother attended two of these workshops with her sister, Debbie. The first session was typical. Janet had some insights about herself and felt completely at rest in a state of bliss after her turn being the breather. Months later, when they returned for a second workshop, things went a bit differently. In the middle of her turn as the breather, Janet sat upright and began speaking an unknown language, at least it sounded like a language, but nothing she had ever heard before. Aside from one year of high school Spanish, Janet had never studied any other language besides English, had never been out of the United States, and had no idea what she was saying—if anything. These words were just flowing out of her. After the holotropic breathwork session ended, Janet went home excited and a little scared about what had happened. She began slowly telling her family and friends about the experience and quickly found that she could still access this ability. Simply by shifting her awareness, she could turn it on, and the language(s) would just pour out. In fact, at the beginning, the languages sometimes seemed to have a mind of their own, spontaneously erupting without an invitation. Eventually, Janet learned how to control the languages, allowing them to come through only when she chose. Of course, at this point, she wasn't even certain that this was a language. It felt like a language, but nobody seemed to recognize it. It was possible that she was just making up sounds that gave the appearance of a language.

It turns out that there are at least a few well-known conditions in which people begin spontaneously speaking another language—or what sounds like a language. *Glossolalia* is the practice of "speaking in tongues," that occurs in certain Pentecostal and charismatic Christian churches. With glossolalia, what is spoken is not recognized as a language and there is generally not any interest or attempt to translate what is said. Instead, this practice is typically seen as a sign of the Holy Spirit taking over the physical body. *Xenoglossy* is the phenomenon of speaking another language of which the person previously had no knowledge. This is obviously rare, and considered controversial. Most xenoglossy cases are connected to hypnotic states or believed to be in connection to some kind of retained memory from a past life. If Janet's experience was glossolalia, it was certainly a very different manifestation than most known cases. First, it wasn't happening in the context of an ecstatic religious ritual. Second, it kept happening spontaneously after the first incident: in the grocery store, while driving her car, or cooking dinner. It didn't quite seem to be xenoglossy either. If Janet was speaking a language, she had no idea what she was saying, so it was not functional in the same way as other reported cases. So what was happening?

The Language(s)

Since the process began, Janet had been recording herself during these language experiences and sending the tapes to professors and researchers from all over the United States. While many of the experts were polite, they had no idea how to help. Other times Janet would receive a response suggesting that she was psychotic or suggesting that this was glossolalia and simply gibberish. Being tenacious, she kept at it and, after four years of searching, eventually found someone who was willing and able to translate these languages. The late Dr. Bernardo Peixoto, an anthropologist at the Smithsonian Institution and a shaman, was originally from the

Urueu-Wau-Wau tribe in Northern Brazil, where he was known as Ipupiara or Ipu. He recognized something in Janet's language and indicated that she was speaking Yanomami, a South American tribal language. This was the confirmation Janet had been looking for. Even though she didn't know what she was saying, she always felt that there was a meaning behind the sounds—that they weren't just nonsense. When her exact words were translated, they generally took the form of prayers and teachings related to honoring Mother Earth. Over time, Ipu translated several tapes and reported that Janet was also sometimes speaking several other South American tribal dialects including Fulnio, Tukano, and Kanamari. All of this information, taken together, suggested that Janet was somehow channeling several people, beings, or entities.

We Meet

I was fascinated by this story and wasted no time in contacting Matt's mom, Janet. We spoke on the phone, and she was very open about her ability but also somewhat reserved, waiting to see if I would judge her or attempt to write off her experiences as something it wasn't. For my part, I was trying to be open to what she described but the scientist in me couldn't help wondering if there wasn't some other, more reasonable explanation. As much as Janet appeared sincere, the idea of channeling South American shamans did seem a bit far out. Aside from my own curiosity, it was probably Janet's authenticity that nudged me toward pursuing this study. Janet was clear from the beginning that she wanted to understand what was happening to her. In fact, while she felt strongly that these experiences were meaningful and real, she was also open to, and interested in, any scientific understanding of this process. Could we learn anything from measuring her brain while she was speaking these languages? Could this provide tangible evidence to support her experience?

I learned that Janet lived in St. Louis, less than two miles from my father and less than two hours from my location at the time. We agreed on a date and time and set up our first of many meetings. My first impression of Janet was that she seemed very "normal," whatever that is. She lived in a cute suburban house that stood out in no way from her neighbors. The interior of her home and the way she was dressed suggested nothing unusual. I'm not sure what I expected, probably more stereotypical eccentricities, but there were none. No flowing robes or excessive jewelry, no burning candles, and no incense.

At our first meeting, Janet's husband, Carl, and her son, Matt (my lab assistant), were both present. Everyone seemed very excited about what we were doing and wanted to be involved. As much as I like Carl and Matt, they were not helpful. They had lots of ideas and suggestions (maybe too many) and inadvertently created some additional pressure for Janet. Suffice it to say, most of our other meetings took place at my home office. In addition to channeling shamans, Janet is also a medium and psychic, which gave us a lot to explore.

The EEG Experiments

Before beginning any of our experiments, we always started with a baseline EEG (electroencephalogram—a test that measures the electrical activity in the brain) recording. This is simply an EEG recording obtained while the person is sitting there doing nothing. It is important to record this data to determine how the brain changes during other states of consciousness. Because we were primarily interested in what was happening when the languages were coming through, it was also important to do a baseline recording while Janet was speaking English. In fact, as more evidence that Janet was very normal in most ways, when asked to talk about anything at all, she immediately began talking about the new countertops they were getting for their kitchen. She wasn't talking about angels or ascended masters or crystals—she was talking about updating her kitchen.

For the baseline recordings, we compared Janet's brain waves to a normative database. Basically, we could look at Janet's brain activity in eyes open and eyes closed conditions, translate the squiggly lines of the raw EEG pattern into specific brain waves (delta, theta, alpha, beta), and then calculate how much of each of these were present. This data can then be compared to a group of control subjects that represent an average segment of the population for different age brackets. This allows us to make direct comparisons and determine if someone has patterns of brain wave activity that are abnormal or atypical in some way.

Brain Waves 101

Before going further, it might be helpful to provide a simplified explanation of what brain waves are and how they relate to states of consciousness. To begin, every cell in our body uses electricity (and chemicals) to communicate with other cells. This electricity exists in a range of frequencies that are measured by how many repetitions there are in a second of time. Fast brain waves have many repetitions while slow brain waves may only have a few.

The 5 Primary Brain Wave Bands

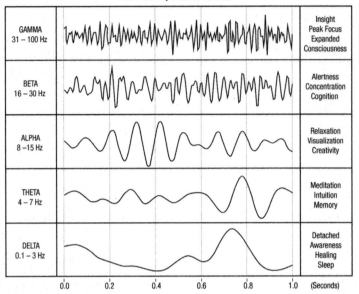

These repetitions are defined by the number of cycles per second (cps), referred to as *hertz* (Hz). Each cycle per second is 1 Hz. Researchers typically put these frequencies into little groups, or clusters, called EEG bands, to simplify the information. These EEG bands are defined by a range of frequencies and given Greek letters as names (delta, theta, alpha, etc.). In general, the frequencies that live together in these clusters tend to have a similar influence on consciousness. Below is a simplified description of the most common EEG bands and their most common associations.

Delta (1–4 Hz): deep sleep, unconscious

Theta (4–8 Hz): twilight state, drowsiness, subconscious, creativity, deep meditation

Alpha (8–12 Hz): relaxed awareness, internal attention

Beta (12–30 Hz): active thinking, attention, problem-solving

Gamma (30 + Hz): higher mental activity, effortless attention

I say that this is a simplification because we actually produce all of these brain waves, everywhere, all of the time.

In addition to the speed of the brain waves, it is also important to consider *how much* of each of these brain waves are present in each moment. The power or amplitude of an EEG signal is measured in *microvolts* (mv). The more microvolts, the more power. So, what is really important is the relative amount of each of these brain waves. When the proportion of brain waves becomes skewed toward one or another brain wave, that leads to a shift in consciousness. For example, when there is a large amount of delta, we are probably asleep. When there is a large amount of alpha we are probably "in our heads" with our eyes closed.

The fact that we can attach numbers to the squiggly lines of an EEG is super helpful and is referred to as a *quantitative EEG* (qEEG). When we can determine how much of each brain wave there is, we can start comparing one person's brain wave patterns to other groups of people, mapping patterns associated with certain mental health concerns, and examining various states of consciousness including meditation, psychedelic states, or psychic states.

Janet's Brain

Compared to a normative database, Janet showed a relative imbalance in her brain waves favoring theta activity. There was also a "hot spot" of beta and high beta in the left and right frontal regions, however, this was likely a result of muscle tension and should not be interpreted. The maps below reflect how the amount of each brain wave observed compared to what is expected. Any shades in the middle of the scale are in the average range. As an area gets darker (black), this indicates that there is more activity in that particular EEG band/location compared to average. As an area shifts to the lighter shades (white), this indicates less activity than expected.

Janet's Baseline Brain Waves Compared to a Normative Database

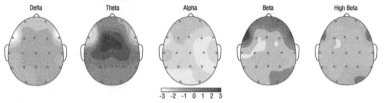

The increased theta activity is located in the frontal lobes, which is interesting as this tends to be a somewhat typical pattern seen in ADHD. This doesn't mean that Janet has ADHD, but what it does suggest is that some of the expected functions of the frontal lobe may be behaving in a less than efficient or expected manner.

One of the primary roles of the frontal lobes is inhibition—the ability to monitor yourself and stop yourself from saying or doing certain things. This is one reason this pattern is often associated with ADHD. Folks with this diagnosis often struggle with impulsivity. Kids with ADHD might blurt out their thoughts or get out of their seat in school without even considering asking for permission. But having excess frontal theta may not be all bad. It is possible that Janet's baseline brain wave patterns help her to access these languages. Because the brain may be less efficient in preventing the expression of thoughts, feelings, and behaviors, it may make it easier for these languages to come through unencumbered. This interpretation fits with Janet's description of what it was like just after the holotropic session where they emerged. She described multiple situations in which the language would just start coming out of her and it would be difficult to contain it. In fact, even when we were working together, Janet would sometimes report that the sensation to speak the language was so strong that it felt like a pressure that was building and needed to be released. Of course, theta activity has also been associated with access to the subconscious and nonordinary states of awareness.

Theta and the Subconscious

Anna Wise, one of the pioneers in the study of brain waves in relation to meditative states and author of *Awakening the Mind: A Guide to Harnessing the Power of Your Brainwaves*, has described theta as the brain wave linked to our subconscious, and consequently our ability to "know" deep information without using the verbal, logical mind (2002). Theta brain waves are often associated with creativity, visualization, and memory, as well as hypnotic states. In fact, theta is the dominant brain wave pattern of young children and may be part of the reason they are nondefensive, imaginative, and open to a sense of magic in the world. It is not uncommon for young children to have imaginary friends, contact with spirits, and a deep connection with the Earth and natural cycles. This is typically observed until around the age of eight, which coincides with a shift in brain wave dominance from a theta rhythm to an alpha rhythm.

Now, this is only what Janet's brain was doing at baseline, pointing to some potential tendencies of the brain that may be related to her abilities. We wanted to know what was happening when she was speaking these languages.

Let the Channeling Begin

After the baseline recordings, I asked Janet to begin speaking the languages. With zero hesitation, she opened her mouth and what sounded like words in a foreign language just poured out. I was trying to focus on the EEG, looking for anything that might be of interest and making sure that the sensors on the EEG cap stayed connected, but to be honest this was very difficult to do while she was speaking. The language is mesmerizing and draws you in. Having heard Janet give demonstrations many, many times in front of various groups, the reaction is always the same. People become entranced and often feel that they know what the message is even though they don't understand the language. In many situations, I have seen people start crying when Janet is speaking. Somehow, there seems to be a truth in the message that transcends a cognitive understanding of what is being said.

When I went back later to review the EEG recordings, there were several times when the record looked very unusual. Rather than the normal clean, squiggly lines, I saw huge excursions, giant spikes, and slow wave activity of tremendous power. Now, normally if I saw something like this, I would assume that there was a bad connection. Maybe one of the electrodes came loose during the session. This would be the obvious interpretation but didn't seem to hold up in this case. Not only was I monitoring the connections during the entire session, but these strange EEG patterns would come and go during the session, looking normal one minute and then jumping off the screen the next. Furthermore, this was not just happening in one electrode, but in several electrodes that were all in the same general vicinity. Having been in the neurofeedback world long enough, I still didn't fully trust that these aberrant patterns were real. I assumed that something was wrong with the electrocap I was using. The next time we met, I made sure to use a different electrocap; one that I had recently tested and took extra care to make sure the connections were 100 percent solid. Even with these extra precautions, the same strange pattern showed up.

Janet's Brain on Channeling

The image below shows a screenshot of what I saw. Normally I look at nineteen separate lines of EEG, but to keep it simple, I reduced it to five. The raw EEG represented by the first, second, and fourth lines on the chart below are what we expect to see. The third and fifth lines are the things that got my attention.

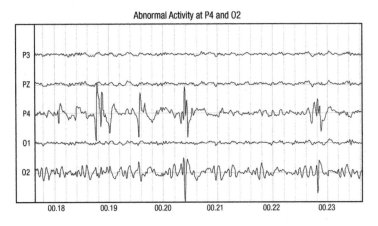

I ran this data through another software algorithm called *LO-RETA* which is an acronym for low-resolution brain electromagnetic tomography. LORETA programs allow us to take EEG data collected from the surface of the head and extrapolate to deeper regions in the brain. Essentially, it takes the information from those nineteen electrodes and estimates where the signals are coming from below the surface. In the end, you get a set of images that look kind of like an MRI, helping to localize the areas of the brain most involved in certain EEG patterns. When I applied this to Janet's wild EEG pattern, it highlighted an area in the back right quadrant of the brain called the *superior parietal lobule*. This hot spot is shown in the image below.

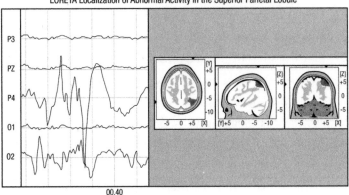

This raises some obvious questions. What is the function of the superior parietal lobule and why is it acting so oddly in this circumstance? Prior to this experiment with Janet, I had not looked in depth at this part of the brain. As soon as I got back to my office, I jumped headfirst into a literature review to understand what this might mean. I didn't have to look long before I found a very interesting study conducted by a colleague of mine from the University of Missouri—what are the chances?

The "God Spot"

I was familiar with Brick Johnstone but didn't know him well as his primary area of study was pretty different from mine. He was heavily involved in studying traumatic brain injury (TBI) and the field of rehabilitation psychology, while I was in clinical and counseling psychology; just different enough that our paths seldom crossed.

In 2008 and again in 2012, Brick studied a group of people with TBIs that had decreased functioning in their right parietal lobe—the same general area that was "acting up" in Janet's study. What they found was that decreased functioning of this region of the brain was consistently associated with reports of increased spiritual transcendence as assessed by the Core Spiritual Experiences Survey (Kass, 2000). This questionnaire asks respondents to indicate if they have ever had certain spiritual experiences, including, "an experience of God's energy or presence," "an experience of a great spiritual figure," "a miraculous event," "an experience of angels or guiding spirits," "an experience of communication with someone who has died," and "an experience with near death or life after death." People with altered functioning of the right parietal lobe tended to endorse more of these items, leading the media to dub this part of the brain, "the God spot." Studies by other researchers seemed to support this idea, finding that brain tumors in the left and right parietal lobes led to increased reports of self-transcendence (Urgesi et al., 2010).

Based on these findings, Brick and a colleague developed a model to explain how this might work (Johnstone and Glass, 2008). Essentially, the right parietal lobe (RPL) is involved in defining and perceiving the self, self-related thoughts, perception of the body, and autobiographical memory, basically understanding the "self" as a separate and discrete entity that is associated with the definition of "me." This region of the brain, in connection with a few others, also seems to be important in distinguishing the self from others. In fact, if you stimulate the right parietal lobe (but not the left), with transcranial magnetic stimulation, you can interfere with this ability (Uddin et al., 2006). We also know that increased activity in the RPL leads to a tendency to focus on the self, which supports the idea that less activity in the RPL leads to less focus on the self or *self-less-ness*. When you are not focusing on your "self," it is easier to connect to something beyond the self, which can be experienced as a mystical or transcendent state. We can write it out like a geometry proof:

If the RPL is involved in defining the self

And the normal functioning of the RPL is disrupted,

Then it is easier to connect to energies beyond the self

Somehow, Janet was able to temporarily disrupt the functioning of her RPL, presumably allowing her to expand her consciousness in a way that allows other forms of consciousness to speak through her. This sounds crazy as I write it, but it is the best explanation for what I have observed.

Why Would Breathwork Cause This?

In her book, *Spirits . . They Are Present,* Janet describes beginning to see and interact with ghosts at the age of five. She also tells numerous stories of premonitions and the ability to "know things" about other people or events. Clearly, Janet already had some psychic and mediumship abilities before her holotropic breathwork experience. Despite these natural abilities, there was something about the HB practice that seems to have created an opening in consciousness, or expanded the crack that was already there. In any case, why would breathing in a deep and rapid manner for an hour cause this to happen?

Holotropic breathwork researcher, James Eyerman, suggests that there are two major elements in the HB process that appear to facilitate access to nonordinary states of consciousness; these include voluntary breathing and music designed to encourage specific emotional and psychological states. Apparently, the neurophysiological effects from these two combined practices leads to a "loosening of emotional control, integrative states, and bonding experiences" (Eyerman, 2014).

Over-breathing

It has been suggested that voluntary over-breathing can cause a significant decrease of blood flow in the brain. This reduced activation could result in a relaxation of the brain's normal inhibitory functions, allowing more subconscious and unconscious processes to emerge (Rhinewine and Williams, 2007). Essentially, by minimizing the "normal" functioning of the brain, aspects of the mind that are normally restrained are set free.

Another theory suggests that the over-breathing or hyperventilation in HB stimulates the *vagus* nerve, leading to a variety of neurochemical changes that may be involved in mystical or trance states. The vagus nerve is the tenth cranial nerve, connecting the brain to virtually every other system in the body. When the vagus nerve is stimulated, it leads to a decrease of sympathetic arousal (the body's stress response) and an increase in parasympathetic tone (the body's relaxation response). In addition, this process can also lead to the release of a variety of brain chemicals that influence mood, motivation, and feelings of connection to others (e.g., oxytocin, dopamine, serotonin, and norepinephrine; Gerbarg and Brown, 2005). Interestingly, it has also been proposed that certain yogic practices involving breathing and chanting also impact the vagus nerve, leading to improvements in anxiety, depression, insomnia, symptoms of PTSD, as well as expanded states of consciousness (Gerbarg and Brown, 2005).

Music

Stanislov Grof created holotropic breathwork after being restricted from using LSD with his psychiatry patients in the 1970s. Grof saw the healing potential in psychedelic medicines and was looking for a nonpharmaceutical way to induce a similar state of consciousness. In his work with clients, Grof noted that the music used in these sessions seemed to directly impact the tone and quality of the psychedelic experience and could be a way to nonverbally support the client's process (Grof, 1980). His music playlists would essentially create a journey for the client, guiding the session through various "moods." In fact, it was observed that the psychedelic session would generally move through phases that occurred in a predictable sequence, including: pre-onset, onset, building towards peak, peak, re-entry, and return. The music was selected to be consistent with these phases. So, during the "pre-onset" phase, the music would be very gentle, quiet, and meditative. During the "peak" phase, music that evokes strong emotional feelings would be used (Bonny and Pahnke, 1972).

In HB sessions, loud, strong, driving music is used right at the beginning to help "activate" the effort involved in over-breathing. More emotionally challenging music is placed in the middle of the experience leading to more soothing, gentle tracks near the end of the session. When used in such an intentional way, the music becomes a primary influence in both psychedelic and HB sessions. This idea appears to be supported by the reports of psychedelic-assisted therapy patients as well.

In a recent study, psychedelic therapy participants were asked to share their impressions of the music used during their session. Seventeen of the nineteen participants (89 percent) reported that their subjective experiences during the session were intensified by the music. Fifteen of the nineteen subjects (79 percent) reported that the music in some way "guided" the journey. The same percentage of respondents (79 percent) reported feeling that the music was supportive, helping them feel grounded and reassured (Kaelen et al., 2018). The reality is that music, by itself, can lead to an altered state of consciousness if used in the right setting with the right intention.

Back to Janet

It is easy to imagine a scenario in which Janet's right parietal lobe (RPL) was already primed to experience more subtle aspects of reality. Remember, she began having both spirit communication and psychic experiences at a very early age. When she engaged in the HB breathing and the blood flow to her brain decreased significantly, it caused the functioning of the RPL to become even more "dysregulated." The boundaries that were typically engaged, keeping a separate identity from other consciousness, disappeared, creating space for this new ability. The neurochemical changes that occurred from stimulation of the vagus nerve created a simultaneous excitation and relaxation. The ramping up of serotonin and dopamine added to the psychedelic effects, while the increase of oxytocin increased feelings of connectedness. The music added color to the experience, guiding, expanding, and supporting it, perhaps orienting her toward some type of shamanic consciousness. In short, the elements of the HB experience may have provided the perfect storm needed to activate this potential.

If Janet, Why Not Me?

Thousands of people, including myself, participate in HB sessions every year without unlocking superpowers. Why is Janet's case different? Maybe it is not different. Perhaps others have experienced an increase in some form of psi ability following HB, we just haven't heard about it. In fact, if Janet had not been so tenacious in seeking answers and if our paths had not crossed, I would be unaware of any such instances. It does not seem like a stretch to imagine that there are others out there that have also had dramatic impacts.

In fact, when I think back to some of my HB experiences and those of friends and colleagues, there are many instances of people reliving parts of their lives, interacting with the "memories" of deceased loved ones, and encountering archetypal entities, angels, spiritual beings, and gods. My professional training has led me to assume that these experiences were psychological in nature, a type of emotional drama, representing aspects of the mind. It had never occurred to me that perhaps these experiences were real. I wonder how my feelings and perceptions about myself would be different if I held a different interpretation of my experience. Rather than viewing myself as someone who "doesn't have this ability," what if I viewed myself as someone who has been able to connect to the spirit world? Would this make it easier to do it again? Would it help open the possibility of developing this ability without HB or other techniques?

Try It Yourself

A Note of Caution

Before we explore strategies for using breathwork to induce an altered state of consciousness, it is important to recognize that this approach is not for everyone. Because it is significantly altering your brain patterns and neurochemistry, these practices are not recommended for anyone who has any sort of neurological instability, such as a history of experiencing delusions or hallucinations, head injury, or seizures. Grof, the creator of HB, has also stated that this practice is contraindicated for anyone dealing with any serious cardiovascular disorder, such as high blood pressure, aneurysms, history of heart attacks, etc. In general, if you have any significant health concerns, and/or questions about whether this is an appropriate practice for you, it is best to consult with your physician. Below, I will discuss several breathwork approaches, some of which are gentler than HB. You can also simply work with music to explore altered states. Finally, I think it is important to engage in altered state experiences with an experienced and trustworthy guide. Because these experiences can lead to the surfacing of challenging memories or emotions, it is best to have a support person with you.

Breathwork

There are many types of breathwork practices that can be used to facilitate a nonordinary state of consciousness, that is, altered state of consciousness or ASC. Holotropic breathwork is probably the most popular and well-known of these techniques. As described above, this process involves rapid, deep breathing for an extended period in combination with specially selected music, and a supportive environment.

You can also induce changes in consciousness by going to the other extreme and slowing the breath, holding the breath, or alternating between these variations. Programs such as SOMA breathwork are based on ancient *pranayama* practices (prana for life energy/breath, yama for control). A component of yoga, pranayama is a means of controlling the breath. SOMA breathwork techniques shift between a rhythmic breathing pattern (inhale two, exhale four) and breath retention. During the breath retention elements, the breather holds their breath at the bottom of the exhalation as long as they can a few times in a row, followed by holding the breath at the top of an inhalation. This practice is a bit gentler than HB but can still result in significant shifts in consciousness. You can try a short version of a SOMA-based meditation by visiting www.psychicmindscience.com.

Using breath to induce an altered state of consciousness or ASC is a common approach used in a variety of meditation and spiritual traditions. As mentioned above, the ancient Indian practice of pranayama includes a wide range of such techniques that can be explored by studying Kundalini Yoga, Siddha Yoga, Sufism, Burmese Buddhism, and certain Taoist meditations. A common approach and a good starting point is the "breath of fire."

Adapted Breath of Fire

Begin by simply watching the breath and settling into your body. Allow the breathing to naturally relax and shift into your belly.

Now begin slowing the inhalation. Imagine breathing in through a straw and extending the breath, making it deeper and longer. Purse the lips and let the air out slowly, as if you are letting air out of a tire. You can also let the air out with an audible sigh, "Ahhhh," activating your vagus nerve.

Continue this extended, long, slow breathing for several minutes.

Now, shift your breath to a circular pattern: as soon as you ex-hale, begin the next inhalation without any pause in between.

Experiment with speeding up the pace of the breath.

Experiment with the intensity and depth of the breath.

Practice for ten minutes (or longer) and then relax the breathing, closing the eyes, and resting, Take a moment of mindfulness to no-tice what is happening with your mind, body, spirit.

Musical Journeys

Specific music, if used in the right set and setting, can lead to sig-nificant nonordinary states of consciousness without the introduc-tion of breathing techniques or psychedelic medicines. Although combining music with these other techniques amplifies the effect, I would encourage you to try a session with music alone. An ideal music journey involves several important elements described below.

- **Identifying a playlist.** Central to the experience is access to music that is designed to shift your consciousness and move you through various stages of self-exploration. I generally rec-ommended that the music selected does not involve vocals. Words tend to shift us into our left brain/analytic mode, rather than dropping us into a deeper state of consciousness. In addi-tion, most people have significant music preferences that may influence their response to the music journey. For example, I don't particularly care for classical music in this context and would likely have a negative reaction if something from that genre showed up during my experience. I recommend ex-ploring a few already created playlists on platforms such as Spotify or Pandora. By typing keywords such as "breathwork," "psychedelic," "psilocybin," and "ketamine" into the search bar, you will find a variety of playlists that have been curated by others interested in the role of music for inner work. If you find a playlist you like, then you are set. You can also save songs that work for you to your own playlist and begin creating something specific for your interests and needs. I have created a sample public playlist on Spotify titled "Psychic Mind Science Music Journey." This is a one-hour list that will give you an idea of the format and the shift in musical styles from the beginning to the middle to the end of the journey. Once you have a playlist identified you can begin planning other aspects of the set and setting for your journey.

- **Setting.** Prepare a space for your journey that will not be disturbed and feels safe and comfortable. You might lie on a bed or create a space on the floor with mats, cushions, and blankets. I recommend wearing a blindfold during the session. The increased darkness promotes inner reflection. Similarly, it is usually a good idea to wear ear-covering headphones, although if you feel like this will be uncomfortable it is fine to have the music playing externally in the room. If you choose not to use headphones, I suggest keeping the volume of the music significantly louder than you would normally. Again, helping block out other sounds from the environment, but also drawing you deeper into the experience. Because these experiences can sometimes be emotional, it might also be helpful to have some tissues nearby as well as water or anything else you might need to help you feel comfortable and be able to stay engaged in the session.

- **Set.** This refers to the "mindset" that you bring to the session. Your mental attitude going into the journey can have a strong impact on the experience itself, so it is best to make sure that you choose the right timing for this work. If you are not feeling well, physically or emotionally, it is best to wait. Once you have established a readiness, then it is important to consider how you will mentally and psychologically approach the session. While it is fine to simply show up and enjoy the ride, it can also be beneficial to enter the session with a specific intention. Why are you doing the journey? What do you hope to get from it? Insight? Understanding? Connection with your Higher Self? Exploration of past traumas? Whatever the intention, see if you can get clear. Be specific. Once you have clearly identified your intention for the session, let it go, trusting that you will receive exactly what you need. Attempt to create a mindset that is accepting and grateful for whatever you receive. If things become emotionally intense during the session, see if you can be curious about the experience and keep breathing. Remember that, at any time, you can move out of the experience by opening your eyes, sitting up, and engaging in activities to help you feel grounded.

- **Integration.** After the session, take some time to reflect on your experience. What will you take away? If you gained a new perspective or understanding about something in your life, how can you turn that into a behavior—making it real? Allow yourself some time to slowly shift back into normative consciousness without feeling rushed or pressured. Continue to explore anything that occurred during your session over the next days and weeks. I recommend keeping a journal of your experiences to explore any insights, thoughts, feelings, symbols, or patterns that emerged during the session. Documenting your experiences in a journal can also provide a powerful tool to track your progress over time.

CHAPTER TWO

THE MIND OF A MEDIUM

JANET MAYER FREQUENTLY TALKED about her work with Forever Family Foundation (FFF) during our research. This is a nonprofit organization created to "further the understanding of Afterlife Science through research and education while providing support and healing for people in grief." The FFF is one of a few organizations that has a process in place to certify mediums. After a medium passes an initial screening, he or she then does a series of five mediumship readings with five different trained sitters. This is not a scientific study with tightly controlled, blinded conditions, but it is a rigorous set of tests that evaluate the accuracy of a medium in situations that resemble how they normally work. I can appreciate this approach as I have found that it is often very difficult for mediums, psychics, and healers to do their thing when the situation is overly controlled, or they are asked to do their work in a way that is significantly different from the way they normally operate.

The sitters for these sessions have been trained in the specific procedures used by FFF and have had no prior contact with the mediums being tested. During the reading, the sitter is not allowed to provide any information to the medium except "yes" or "no" type responses to any questions the medium might ask (they can say things like, "I understand" or "I don't know," etc.). After the reading, the sitter provides a score for each of the pieces of information brought forward by the medium, rating the accuracy and specificity. The scores for all five readings are analyzed through several different statistical methods and only those scoring above a certain level pass the test. This is an intense procedure and results in only a very few mediums meeting the criterion. In fact, at the time of this writing, FFF had only certified twenty-seven mediums since beginning the process in 2005.

Janet discussed our experiments with FFF founders, Bob and Phran Ginsberg, who invited us to be guests on their radio show, *Signs of Life*. After the show, we were invited to present our research at their annual conference. At first, I wasn't sure if this was a good career move or not. Did I really want this listed on my curriculum vitae? I had been entertaining the idea of looking for a faculty position rather than working in the Student Health Center. Frankly, working with mediums and psychics is generally not held in high regard in academic circles. On the other hand, it sounded fun, and the foundation was offering to pay for my flight and hotel—and the conference was in San Diego— so why not?

I had never heard of Forever Family Foundation before this and wasn't sure what to expect. I expected that there would be about thirty to fifty attendees consisting of people mostly interested in receiving a mediumship reading with minimal interest in the scientific aspects of this work. I quickly found out that my assumptions were entirely inaccurate. The conference had several hundred attendees, was totally professional, and included a range of presenters including some extremely gifted mediums, researchers, and psi investigators. In fact, given my limited knowledge of the field, I felt like the least qualified person to be there speaking on this subject.

The first night of the FFF Conference featured Kim Russo, host of the television show, *The Haunting of...* doing what's called a "gallery reading." This involves conducting a series of brief mediumship readings for the people in the audience. She was impressive and smooth as she identified personal and specific information that she could not have possibly known beforehand. This was the first group reading I had ever seen, and my skeptical brain was watching carefully for any signs of "cold reading techniques." Not that I knew a lot about cold reading, but I am trained as a psychologist. I can see when people are using subtle cues to manipulate or influence others. Cold reading refers to a set of strategies used by would-be mediums to extract personal information from sitters that they can later use in the reading. Basically the "medium" asks questions or makes statements with a high probability of getting an affirmative response. Often this involves making broad statements and using any cues from the sitter based on their age, clothing, or appearance that might get an affirmative response. This often leads to the sitter providing additional information, either verbally or nonverbally, which can be used to guess your way into something that looks and feels like an accurate reading. Of course, the implication is that cold reading techniques are used by charlatans to dupe an unsuspecting audience.

And to be fair, I did observe some of these techniques being used during the weekend although there may be other explanations besides malicious intent. For instance, it seemed relatively common for a medium to approach a sitter with a statement like, "I am getting a person whose name begins with a J or maybe a G." Now, I don't know how the spirit world works and perhaps it is difficult to extract super-specific information, like someone's name. At the same time, this kind of thing seemed like fishing to me and made me a little suspect.

Kim Russo's reading had none of that. In fact, Kim, and all of the other mediums I observed over the weekend, were very careful to instruct the sitters to only answer with an acknowledgment if they said something that seemed accurate. I was impressed.

Over the course of the weekend, I was introduced to dozens of mediums and researchers, several of whom became good friends and collaborators. This is where I met bestselling author and medium Laura Lynne Jackson. Janet introduced me to her and insisted that I would love her. She wasn't wrong. Laura is clever, warm, compassionate, and really funny. She knew of the work I had been doing with Janet and volunteered to have her brain mapped. Because of the schedule for the weekend, we did not have a chance to do the EEG recording until the end of the final day of the conference.

We were in a small conference room, just off the main ballroom, working at a large table covered with a standard hotel tablecloth, there was a fake plant in the corner, and nothing on the walls. We decided it would be a good idea to have her conduct a "live" reading while recording her EEG. Because my mom was along for the trip, we initially thought we would use her for the reading. However, as the weekend developed, I had a strong feeling that I should ask Laura to read me instead. The scientist part of me felt uncomfortable about this as I was supposed to be objective and separate from the subjects. Despite my reservations, I could not ignore my intuition and asked Laura to read for me during the experiment. Boy, am I glad I did.

Laura's process of engaging in a reading typically begins with what she describes as the psychic portion, which she perceives in her left visual field. She described it as "seeing" something (called *clairvoyance*), like on a television screen, where information related to the person will show up. This portion of the reading lasted for approximately thirty minutes, which she later indicated was much longer than typical. While some of the information she provided was somewhat general, much of what she offered was very specific and accurate. Laura told me that she was seeing my core aura as green and suggested that my mission was to keep changing and growing, to be adventurous, to travel more. She also predicted that there was a move in my semi-near future. She speculated that I may be getting a job offer. In fact, my wife and I had our house on the market, and I had been positioning myself for jobs in several other cities. One year after this reading, we moved from Missouri to Oregon, and I started the NeuroMeditation Institute.

Laura noted an orange tone on the left of my core aura, which she related to an appreciation of art. She then mentioned music and specifically a trumpet. I absolutely love music and have dabbled with playing the guitar and drums. One of my life goals is to be good enough at an instrument to play with a band. Also, the very first instrument I ever played, in fifth grade, was the trumpet.

She told me I would go to Hawaii and then return in two to three years. She also said that I would work with another medium, "maybe Joanne Gerber," and that I would go to New York to continue my scientific investigation of mediums. She told me I would do some work in New Jersey. At the time, I didn't know where to put any of this information, but guess what? It all came true. I ended up going to Hawaii three times over a four-year period. I eventually worked with Joanne Gerber. I went to New York to do more testing with Forever Family Foundation, and I traveled to New Jersey to do research with arhatic yoga and pranic healing.

Laura accurately described my wife, Erika, indicating that she had a good vocabulary, was smart, but not as driven as I am. She said that I will "throw things wide open" and Erika will be supportive. She also mentioned that Erika always wears black and makes the bed every day. Again, while none of this is mind-blowing, it was all accurate. In our relationship, we joke that Erika is all earth and water, keeping me grounded while I explore and create. Erika is extremely smart and prides herself on her use of language, her love of words. She likes to have things tidy and organized, keeping her environment very Zen-like, and half her wardrobe is black.

Laura knew I had two children. She told me they were a boy and a girl (correct), that the girl was the oldest (also correct), and then went on to describe their personalities very accurately, which was interesting because they are very different from each other. Laura went on and on, giving accurate information about my mother, my father, my grandparents, and my siblings.

At some point in the reading, Laura switched to more of a mediumship reading, which she perceives in her right visual field. Laura connected to several relatives and friends of mine who had passed, but the one that really got me was my Grandpa Brunetti. As Laura was tuning in, she said, "I'm getting a grandfather on your mother's side." Okay, sure, that is an easy guess given my age. You can bet my grandfather has probably passed, but then she said, "His name is Giuseppe." What? How in the world could she have known that? Who would just guess the name Giuseppe? My very English last name, Tarrant, offered no hint that I have some Italian heritage. This is the kind of information that is termed "evidential," meaning it is precise and specific.

While my grandpa was not a great parent, we had a special relationship. He lived on twenty-two acres in rural Missouri with some woods, a pond stocked with fish, and a vegetable garden. He loved it when I would visit and always made me feel special. He would take me fishing or catching crawdads. We would go to the local pub, and he would buy me an orange soda, himself a beer, and I would sit and listen to him gossip with the other customers. The thing I remember most about him was his huge, ridiculous laugh. He most likely had ADHD and was constantly involved in some kind of crazy idea, looking at ways to entertain himself. He was a prankster and would do things like chase us around the house with his dentures popped out of his mouth, or bring out a block of Limburger cheese and wait for us to gag from the smell. I once got in trouble in kindergarten for singing a song that he taught me: "I got a gal that lives on the hill. She won't do it, but her sister will." To him, this was gold.

Obviously, Laura did not know any of this but gave a dead-accurate description of his sense of humor, stating that he was being inappropriate. She described his joy in gardening and fishing, his career in the Army, his joy in fixing things—even though he wasn't very good at it, and his religious beliefs. Laura added, "There's something funny about dentures." She told me that he was "happy as hell when he died and figured out he was not really dead." Of course, there were other parts to the reading, parts where he apologized for how he treated my mom and expressing his pride in me and what I had accomplished.

After fifty-five minutes of nonstop EEG recording and furious writing on my part, my grandfather told Laura to wrap it up because we were running over time. He also told her we would finish at 7:13. Laura offered a few more pieces of information, started to "close the doors" on the session, breaking her connection with the spirits, but first asked me if I had anything I wanted to ask. I apologized to my grandfather for not attending his funeral, which is one of my biggest regrets in life. Grandpa basically indicated that there was nothing to apologize for, it was in the past, "water under the bridge." When this interaction concluded, Laura wrapped it up, and I stopped recording. Just as I got up to take off the EEG cap and clean up, I checked the time on my computer. It was 7:13.

This was more than I expected. So much of her information was so accurate and specific that I could not ignore it. I had tears in my eyes by the end of the reading and felt a glimpse of how healing this process can be for those that have lost someone close to them. Having a sense that our loved ones are still "out there," that they are still paying attention and trying to help from the other side makes the loss more bearable. Still, the scientist part of me could not help but consider alternatives to the notion that Laura was in contact with my deceased grandfather. I had to accept that Laura was able to access information that she could not have known from traditional means, but couldn't this be a psychic ability? Isn't it possible that Laura was actually reading something about me or my memory, hopes, dreams, and regrets, and this information was being interpreted as coming from another entity? In favor of this hypothesis, most of the information that came through during the reading, whether psychic or mediumistic, related to me very directly and matched my perceptions. My grandfather behaved as I remembered him. He forgave my absence from his funeral, which is what I wanted. He acted with more wisdom and maturity than when alive, playing the role of a guide, which is what I would like to believe. Isn't it simpler to assume that Laura was "reading my mind" rather than communicating with the dead?

On the other hand, he said things during the reading that did not relate to me directly and for which I had no knowledge. For example, he apologized to my mother for something he did during her twenties. He said he treated her unfairly and didn't "see things clearly" with her at that time. Then, there are the results of the brain imaging. Laura is a bit unique among the mediums I have met in that she experiences a very clear distinction between engaging in psychic readings and mediumship readings. For Laura, she experiences these different states on different sides of her visual field. This suggests that these two states should show up differently in her brain.

Visual Processing in the Brain

Without getting too complicated, the right and left visual fields are divided in each eyeball, such that the right side of each eye is tuned to the left visual field and the left side of each eye is oriented toward the right visual field. When you track this through the brain to the visual processing centers in the occipital lobe, it also demonstrates this crossover process. So, information viewed in the right visual field should be reflected in activity in the left occipital lobe, whereas information viewed in the left visual field should be represented in the right occipital lobe.

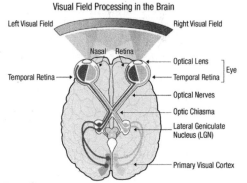

Visual Field Processing in the Brain

Laura's Psychic Brain

To begin, I compared a sample of Laura's EEG from the psychic reading to an EEG sample when Laura was just having a normal conversation. By comparing these two conditions, I could feel more confident that any differences in the data were not simply due to talking in one while remaining silent in the other. Below is a series of brain maps for different brainwave frequencies showing the difference in brain activity between having a regular conversation and the psychic-only portion of the reading. Areas of the brain with darker shading indicate a significant increase of activity during the psychic reading. Lighter areas indicate a significant decrease of activity during the psychic reading.

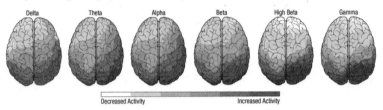

During the psychic portion of the reading, Laura's brain wave activity clearly showed increased activation across all EEG bands at the back of the head. Specifically, this activity appears to be leaning toward the right. Clearly this suggests that Laura is processing visual information differently during a psychic reading than during normal conversation. Interestingly, the strongest changes occurred on the right side of the occipital lobe, which is connected to the left visual field, exactly what Laura described when explaining her process. This blew my mind. Objectively, nothing changed in the environment. Laura's physical eyes were seeing the same boring conference room during the conversation as they were during the psychic reading. Whatever Laura was seeing in her mind's eye was being processed in the brain in the same areas where the brain processes visual information from the physical world. Now, we can't necessarily say that this is proof that Laura's psychic skills are real, but it does provide some compelling evidence.

Laura's Medium Brain

Next, I compared Laura's EEG from the mediumship reading to the conversation-only data. I wondered how this would look in comparison to the psychic data. During the mediumship reading, there was an increase of slow brain wave activity (delta, theta, and alpha) in the occipital lobe, focused on the left side, consistent with information in the right visual field. While this finding fits with what we might have predicted based on Laura's description of her process and the results of the psychic testing, there were also some interesting differences. While Laura was giving the psychic reading, her brain showed a strong increase of activity across all EEG bands in the right occipital lobe, with the strongest increase of activity in the "fast" brain waves (high beta and gamma). During the medium-ship reading, the increase of activity seen in the left visual field was only observed in the "slow" brain waves, with the fast brain waves showing only a minimal change. This suggests that there are additional differences in brain activity between psychic and mediumship readings beyond which visual field they appear in.

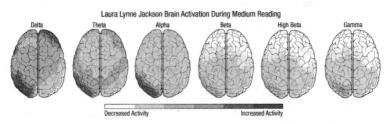

Laura Lynne Jackson Brain Activation During Medium Reading

Delta   Theta   Alpha   Beta   High Beta   Gamma

Decreased Activity   Increased Activity

The other interesting finding with the mediumship data relates to the significant decrease in fast activity in the frontal lobes. While we saw a decrease in fast activity in the frontal lobes during the psychic reading, it was even more pronounced in the mediumship reading, with the entire frontal lobe shutting down.

Psychic Versus Mediumship

While these initial analyses gave me plenty to think about, I wanted to see a direct comparison of the psychic and mediumship readings. Rather than comparing the psychic/mediumship recordings to a conversation, I compared them to each other. For these images, darker colors indicate areas where the mediumship condition had more activation than the psychic reading, which was only in the slow brain waves located in frontal and left occipital regions. Lighter colors indicate areas where the medium condition had less activation than the psychic condition. These changes were focused on the fast brain waves in the frontal lobes and the right occipital region. These findings clearly indicate that during a mediumship reading the brain was significantly less active than during a psychic reading (more slow activity and less fast activity).

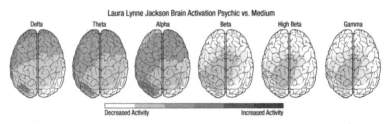

Laura Lynne Jackson Brain Activation Psychic vs. Medium

Delta   Theta   Alpha   Beta   High Beta   Gamma

Decreased Activity   Increased Activity

Coherence

When examining brain waves, connectivity between different locations can be measured with a metric called coherence. This type of analysis is essentially giving us information about the amount of shared activity between brain locations—how much or how little they are working together. If areas of the brain become too connected (*hypercoherent*) or disconnected (*hypocoherent*), this can cause problems as the areas of the brain are not engaged with each other in optimal ways. Changes in coherence can also indicate shifts in consciousness. When I examined changes in Laura's brain during both psychic and mediumship readings, there was a clear movement toward less coherence, or disconnection, between regions. This shift is reflected in the image below. Lines between sensor locations indicate areas of hypocoherence; the darker the line, the stronger the pattern.

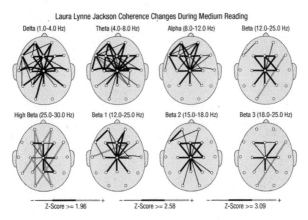

Laura Lynne Jackson Coherence Changes During Medium Reading

Why would areas of the brain become less connected? Perhaps this tendency is the way that Laura is able to move her ego aside so she can clearly receive information outside of her normal waking state. This idea seemed to make sense as the typical communication patterns in the brain are likely designed to process information and thoughts to support our day-to-day functioning. Talking to dead people is obviously a bit different and may require the shutting down of those normal communication patterns.

Interestingly, research from UCLA may indirectly support this idea. The authors of this study were examining what happens in the brain when we slip into unconsciousness. According to one of the authors, "When we lose consciousness, the communication among areas of the brain becomes extremely inefficient, as if suddenly each area of the brain became very distant from every other, making it difficult for information to travel from one place to another"

(*Science Daily*, 2013). Because consciousness is generally understood as being fully alert and aware of the construct of who you are as a person, this seems to fit with our findings. Many mediums describe the process of receiving information as getting the ego out of the way and becoming less attached to the idea of self. It is a process of letting go. So, in addition to increasing or decreasing certain brain waves, it also makes sense that mediums may (consciously or unconsciously) shift how areas of the brain are communicating (or not communicating) with each other.

Laura's Brain Summary

First, the brain pattern observed during a psychic reading was different from what was observed during a mediumship reading, suggesting that these two skills require somewhat different types of brain activation/deactivation (we'll explore the psychic brain in more detail in Chapter 4). In Laura's case, there was more activation in the brain during a psychic reading, especially in the visual cortex. During the mediumship reading the only brain waves that increased were slow brain waves, suggesting a decrease of activation, again in the visual cortex. Perhaps even more interesting was an examination of connectivity between brain areas while Laura was engaged in psychic and medium processes. Across all brain wave frequencies, there was a tendency for the brain to decrease communication between regions, potentially signaling a shift in consciousness toward a more open state. It was amazing to see such dramatic changes in the brain that mapped so clearly to the change in function we might expect. Of course, this is data from one medium. I had no way of knowing if this was typical or something unique to Laura. Once I returned from the conference, I decided to expand my study. I wanted to know if other mediums showed changes in their visual cortex and/or coherence patterns like what was observed with Laura. Now I just needed more mediums who were willing to have their brains mapped.

Other Mediums

If you are going to test the brains of mediums, it seems important to make sure they are "legit." To put it plainly, I would like to know if there is any evidence that the medium I am testing actually possesses a significant level of mediumship ability. How do you test this? Fortunately, I don't have to—organizations such as Forever Family Foundation have elaborate testing procedures in place for just this thing. Mediums who can consistently provide evidential readings under semi-controlled conditions, with trained sitters, as well as passing all other screening criterion can be considered certified. In short, the mediums certified have proven themselves—they are the real deal. These were the mediums I was interested in working with.

Given that Laura Lynne was certified with FFF and also Windbridge Research Center in Arizona, which studies "dying, death, and what comes next," I was off to a good start, but I needed a few more participants. I reached out to Bob Ginsberg, the cofounder and Vice President of FFF, and discussed my interest. He invited me and several of the FFF-certified mediums to his house for a long weekend of mediumship testing. Over the course of two days, I was able to work with three additional mediums: Joanne Gerber, Angelina Diana, and Rebecca Anne LoCicero. I was curious if the data would show any similarities or differences in their brain patterns when they were doing their thing. Specifically, how did their brain patterns look next to Laura's? Did they also show changes in the visual cortex? What about the frontal lobes? Any dramatic shifts in coherence?

For each medium, I compared their EEG data while having a regular conversation to their EEG data while giving a mediumship reading. I began by first just looking at the amount of EEG power in each band (delta, theta, alpha, etc.) in the four primary lobes of the brain: frontal, temporal, parietal, and occipital.

To keep it simple, I will refer to delta, theta, and alpha as slow brain waves, and beta, high beta, and gamma as fast brain waves. The image below provides a summary of my findings across all four mediums. The direction of the arrows indicates if the activity increased or decreased, while the shading of the arrow represents if this was in relation to slow or fast brain waves. There are more than four arrows as some mediums demonstrated more than one pattern.

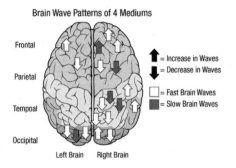

Brain Wave Patterns of 4 Mediums

Clearly there were some interesting patterns here. First, all mediums showed significant changes in the frontal lobes. While there was no real consistency in how they shifted their brain (e.g., some increased fast activity, while others decreased it), there was a definite bias toward the right frontal lobe. Because this area of the brain is intimately connected to various aspects of attention, it suggests that the mediums are paying attention differently when in a mediumship state. Three of the four mediums also showed changes in the parietal lobe, again with no consistent pattern regarding the brain waves, but again with a significant bias toward the right side. In fact, none of them showed changes in the left parietal lobe. Interestingly, the right parietal lobe is the same region of the brain that was "acting strangely" when Janet was channeling other languages (see Chapter 1). This suggests that shifting the normal functioning of this part of the brain is important for many mediums to move beyond ordinary self-boundaries and connect with something outside of themselves. Three of the four mediums showed significant changes in the occipital lobe, which is the primary area where visual processing occurs in the brain. Again, there was no pattern in terms of specific brain wave changes. For this specific region of the brain there was also no clear preference for one hemisphere or the other. I suspect this activity is related to the process of "seeing" (clairvoyance) on the other side. If this is true, it makes sense that mediums who receive more information through visual senses are also the same mediums who show significant changes in this region of the brain.

Coherence Comparisons

For this analysis, I looked at changes in coherence among all nineteen electrodes for each of the four primary EEG bands (delta, theta, alpha, beta), in each hemisphere and across hemispheres, resulting in a whopping 256 possible combinations. Because coherence is a constantly changing dynamic, it is a little difficult to know if any changes observed are relevant; they could be just normal fluctuations. To determine which (if any) changes were meaningful I entered the data into the NeuroGuide software analysis program. This program allowed me to determine how many of the coherence changes were statistically significant, and unlikely to have occurred by chance. In the chart below, you can see the percentage of coherence measures that changed significantly for each medium in the left and right hemispheres, as well as across hemispheres.

Percent of Significant Coherence Changes During Mediumship

	LEFT HEMISPHERE	RIGHT HEMISPHERE	CROSS-HEMISPHERE
Laura	5.5%	17.2%	9.4%
Angelina	5.5%	18.8%	28.1%
Joanne	8.6%	12.5%	9.4%
Rebecca	37.5%	40.6%	56.3%

This comparison shows that changes in coherence are common with mediumship and tend to be more heavily focused in the right hemisphere (again). Angelina and Laura showed a much higher percentage of coherence changes in the right hemisphere compared to the left hemisphere. In both cases, the coherence changes on the right side primarily involved decreasing connectivity. Two of the mediums (Angelina and Rebecca) also showed a relatively high percentage of coherence changes across the hemispheres. This suggests that the way the right and left sides of the brain are talking to each other shifted significantly during the mediumship reading. Finally, the observed changes clearly favored decreasing coherence or a combination of increasing certain areas of connectivity and decreasing others. At this time, it is unclear if the degree of change seen in coherence patterns is related to the accuracy or clarity of information received. While this is certainly possible, it is also possible that less change is related to stronger mediumship. For example, it could be argued that mediums showing less change are more naturally "tuned in" to the other side and require fewer brain changes to access information from the other side.

Taken together, these findings indicate some commonalities and differences in how the brain changes its functionality to facilitate access to information that is not normally accessible. Common patterns include changes in specific brain regions including the frontal lobe (especially right side), the right parietal lobe (RPL), and the occipital lobe. In addition, each medium demonstrated numerous significant changes in brain communication patterns. Whether there was an increase or a decrease in coherence, these changes imply a shift away from typical communication, potentially allowing certain parts of the brain to operate more independently and others to become more closely linked. In any event, it appears that the most common change in coherence is to decrease connectivity. As mentioned earlier, I suspect this allows the medium to get out of their normative consciousness, to essentially break down or let go of the normal way of thinking about themselves and the rest of the world, which is supported by research with psychedelics and meditation.

Coherence and the "Ego"

Barnett and colleagues (2020) examined changes in functional connectivity in the brain while participants were under the influence of LSD, psilocybin (aka magic mushrooms), or low dose ketamine. With all three medicines, the researchers observed a general decrease in functional connectivity throughout the brain. They suggest that this shift in brain dynamics is related to "a breakdown in patterns of functional organization or information flow in the brain." Another study with psychedelics found that reductions in connectivity within the default mode network (DMN) were linked to experiences of *ego dissolution*, that is, a breakdown in one's sense of self (Smigielski, L. et al., 2019). Interestingly, research into the brain on meditation has also found that decreased coherence is a common pattern. In a study examining expert practitioners from five different meditation traditions (thirteen Tibetan Buddhists, fifteen qigong, fourteen Sahaja Yoga, fourteen Ananda Marga Yoga, and fifteen Zen), Lehmann and colleagues found that all of them showed decreased connectivity during meditation compared to a resting state (2012). The authors concluded that, "The globally reduced functional interdependence between brain regions in meditation suggests that interaction between the self process functions is minimized, thereby leading to the subjective experience of noninvolvement, detachment, and letting go, as well as of all-oneness and dissolution of ego borders during meditation." This suggests that meditation may be the perfect training tool for developing a mediumship-attuned brain.

Try It Yourself

As you work to develop and grow your own mediumship skills, it may be helpful to also practice quieting your mind, decreasing coherence, and shifting out of your normal, waking, everyday consciousness into something that is more expansive. Based on what we have seen with FFF-certified mediums, it is important, if not essential, to quiet the monkey mind in order to hear or see or feel the more subtle energies around you.

As reported above, meditation is one method to help develop the ability to quiet the mind. If you have a busy or anxious brain, you might believe that quieting the internal chatter is next to impossible. In fact, if you are like many people, you may have tried to meditate in the past, found your mind way too busy, and determined that you are simply not capable of meditation. Fortunately, there are a few tricks that can help even the busiest brain start to settle down.

Quiet Mind Neuromeditation

This approach is about decreasing activation of the default mode network (DMN). The DMN is a cluster of brain regions working together to create an identity. This network in the brain is normally extremely active, attempting to make sense out of the world by creating stories. Essentially, the DMN takes in information from the environment and then filters it through your life experiences, beliefs, values, prejudices, and cognitive distortions to create a story about yourself and the world. This process is so constant and automatic, we normally have no idea it is happening. This is the storyteller. This is the creator of internal narratives. When we learn to reduce the activity of the DMN and decrease connectivity within the DMN, the internal world becomes much more manageable. This is essentially what happens in meditative practices such as Zen and Transcendental Meditation (TM). Of course, shutting down this network can be quite challenging. Luckily, there are some very concrete strategies you can use to help find this state of inner quietude.

Relax. It may be obvious, but we are a mind/body. What you do to the mind affects the body and vice versa. If the body is tense, stressed, or otherwise unhappy, it is a near guarantee that the mind will also be tense, stressed, or otherwise unhappy. For this reason, when we teach the quiet mind style of meditation at the Neuro-Meditation Institute, we usually spend the first few sessions simply learning to relax. You might try listening to a body scan or progressive muscle relaxation guided meditation, engaging in some light stretching or yoga, or slowing the breathing with a focus on the exhalation (for example, inhale to three, exhale to six).

Give your brain something to pay attention to. At first glance, this advice may sound counterintuitive. If you are paying attention to something, won't that activate the brain rather than quiet it down? Not necessarily. It depends on what you are paying attention to. If you direct your attention to the absence of something—to space, distance, or volume—the brain tends to settle down. For example, right now, place your hand on your lap. Can you notice and feel the distance between your index finger and your thumb? Don't try to think about it, just notice what happens when you put your full awareness into the space.

By giving the mind something to do, you make it much more likely that it will behave itself. The darkness behind the eyelids meditation is a variation on this theme.

Darkness Behind the Eyelids Meditation

Find a comfortable position. Get as relaxed as you can without falling asleep.

Allowing the body to feel completely supported, let go of any and all bodily tension.

Allow the tongue to relax, allow the eyes to relax.

Allow the breathing to become slow, easy, and gentle.

Now, draw your attention to the darkness behind your eyelids.

Allow your mind to become completely absorbed with this vast emptiness.

Continue to breathe, long and slow.

If fleeting images or words move through the mind, simply let them move through without attachment.

Breathe into the space, allowing it to grow naturally and easily.

Rest in the space.*

Begin with focus. Another approach is based on Transcendental Meditation (TM) practices. Rather than focus on the absence of something, TM instructors begin by asking you to put your attention on one thing: a mantra. The idea is to simply repeat a word or set of words (traditionally given to you by your guru). This is a focus style of meditation and helps to stabilize the mind. As the mind settles and the attention begins to relax, the idea is to allow the mantra

A recorded version of this meditation is available at www.psychic-mindscience.com.

to eventually fall away, leaving the mind in a spacious awareness. You begin with attention on one thing and then take away the one thing, leaving your attention on *no thing*. A technique that I like that fits into this category invites you to take turns paying attention to one thing (a sound) and *no thing* (the space between sounds).

The Space Between Sounds

Find a comfortable position that helps you feel alert and focused yet relaxed.

Spend a few minutes slowing the breath and relaxing the body.

When you feel settled, take a deep breath in and exhale with the sound *om*. This can be chanted out loud or silently in your mind.

As you inhale, focus on the space between the oms, the temporary silence. Meditate on the stillness between the two oms and let go . . .

Continue this practice for at least five minutes.

This practice can also be done while listening to someone else chant om I personally like the recording titled, Om: The Sound of Stillness, *by Master Choa Kok Sui.*

CHAPTER THREE

THE GANZFELD
EXPERIMENTS

SEVERAL YEARS AGO my sister, Nikki, who shares many of my quirky interests, suggested I check out the TV docuseries *Hellier*. The series follows a group of paranormal researchers as they explore reports of "goblins" terrorizing a doctor in rural Kentucky. Turns out, the general area of the goblin encounters was the same area that is known for the Mothman sightings, lots of UFO activity, random paranormal events, and reports of cults gathering in the nearby woods and caves. The story goes from weird to weirder (as you can imagine) as they follow various leads attempting to figure out what is going on in this area. As the cast of the show have all been involved in ghost hunting for some time, this influence crept into their approach. In one episode, the team uses something called a Ganzfeld technique in a cave to communicate with any spirits that might be available and connected to their case.

The Ganzfeld technique as shown in the docuseries involves cutting a ping-pong ball in half and taping each half over one of a person's eyes and then shining a red light on their face. This allows the eyes to be open but creates a homogeneous visual field. Basically, the eyes are receiving input, but there is nothing there to process. This technique is often used in conjunction with specific "white noise" input. White noise is sound that includes all of the frequencies that humans can hear, played with equal intensity. This produces kind of a *shhhh* sound similar to radio static. This type of noise is often used as background to help people get to sleep or to block out other sounds in the environment. At the NeuroMeditation Institute, we have several white-noise machines that sit outside of our offices to make it difficult to hear what people are saying on the inside, thus protecting privacy.

This combination of Ganzfeld light and white noise (and variations on this approach) has been used for decades as a technique to enhance latent psychic abilities. It is almost like a sensory deprivation experience, except rather than taking away all sensory stimulation, you overload the senses with a constant stream of input that has no inherent meaning.

In this *Hellier* episode, Greg Newkirk, the person using this technique, reported that he almost immediately began hearing things as soon as the technique was started. As the session continued, Greg started reporting things he was seeing in his mind. He described a ring of trees, moving past environments like he was on a train, seeing a face with large blue eyes, and flashes of light behind treetops. He also described feeling like his body was vibrating, feeling cold as if the temperature had suddenly dropped, and a sense of presence. After nineteen minutes, the white noise suddenly stopped because the audio was buffering. As soon as the sound stopped,

Greg reported that the images and perceptions also stopped.

On the surface this sure looks (and sounds) like the addition of the technology helped facilitate Greg's ability to tune in. Now, we have no idea what he was tuning in to. Could it have been his imagination? Sure. Could it have been some kind of hallucination? Yup. Could he also have been connecting to some unknown energies that were showing him these images, perhaps as a way to communicate or frighten him? Why not? The truth is, we have no idea where those images and sensations came from or why. I was unable to find any research out there looking at the Ganzfeld technique in relation to spirit communication (although there are lots of data related to telepathy). I wanted to know what would happen if people claiming to have some ability to communicate with spirits tried this approach.

A. J. and Damascus

I called up my friend A. J. Jamrose to see if she would be interested in doing some brain-wave experiments. She has a long history of interest in psi phenomenon and has worked with a variety of teachers in the past to enhance her own psychic abilities, including communicating with her spirit guides. A. J. has also done tarot card readings professionally and is now a full-time (and amazing) tattoo artist. She is what my dad would call "witchy," someone or something that seems to have a power or connection to something bigger than this material universe. If you had a dream about someone you had not seen in a long time and then received a message from that person the next morning, for example, that would qualify as witchy. A.J. agreed to participate and ever since has been a guinea pig for a lot of the techniques in this book.

Rather than using ping-pong balls, we used a set of eyeglasses with lights built into the rims where the lenses would normally be.

The glasses are connected to either a small controller box or to the computer, depending on which version you are using. Built-in programs in the software allow you to choose from a variety of settings that will cause the lights in the glasses to flicker on and off at specific frequencies. This approach is generally referred to as light-and-sound stimulation or *audio-visual entrainment* (AVE). The idea is that by providing the brain with a consistent repetitive signal, it will follow along (entrainment). For example, if I chose an alpha program, the lights might flicker on and off ten times a second. The brain responds to each of these signals, producing a 10-cycles-per-second brain wave (10 Hz), which is in the alpha frequency band. So, using this technology, you can influence the brain by "pushing" it toward certain brain wave patterns.

These devices and their specific programs are generally used to induce a state of relaxation, to help with sleep, to increase feelings of alertness, or to improve mood. The main company we work with, Mind Alive, has a Ganzfeld setting in their software. I can choose this setting and the lights just stay on without a flicker, then the participant can choose what color light she wants to use. For our experiment, we decided to stick to tradition and use red light.

We also decided to use a variation of white noise. Whereas white noise contains all the frequencies the human ear can hear, other "colors" of noise, like pink or brown noise, filter out certain frequencies, leaving you with a more restricted range of audio frequencies. This whole thing may need a bit more explanation. Obviously, sound doesn't have a color, so what the heck am I talking about?

Basically, these descriptions of sound borrow terminology from the study of light and apply it to sound. White light contains all the colors. When you send a beam of white light through a prism, you get a rainbow on the other side (think of Pink Floyd's *Dark Side of the Moon* album cover). Each of those colors has a different frequency. Red light, for example, is a faster part of the wavelength spectrum (approximately 625–740 nanometers [nm]), whereas blue is on the slower end of the spectrum (approximately 380–440 nm). In a similar manner to light, sound has a variety of frequencies that can be put into little clusters based on the speed of the sound waves. These clusters are then given color names to convey their relative speed. With the right software program, it is possible to filter the noise to emphasize different parts of the spectrum.

For our experiment, we used an app called myNoise to access a Colored Noise Generator. This program allows you to determine the intensity of each "color" of noise in the signal. Rather than use a straight white noise sound, we decided to try something a bit different. We emphasized both the lowest and highest frequencies (brown and violet). The decision to do this was part intuition and part experimentation. Because we were hoping the sound would shift a person into a state of consciousness that is relaxed but engaged at a high level, it kind of made sense to try the slowest and highest frequencies together. We also tested this in a previous session where we manipulated the sound colors until it "felt right." This exploration and experimentation of frequencies turned out to be an important part of the process.

I didn't offer a lot of direction for A. J. in this session. I didn't want to set up any expectations that could lead to her subconsciously creating an experience that would be interesting to the study. Instead, I simply told her that this was an experiment, and I was curious to see what might happen, to see if it shifted her consciousness in any way. She told me that her intention for the session was to connect to the spirit world and ask for guidance on how to develop her visualization skills. I invited her to verbally report what was happening if she felt like it. I also let her know it was okay just to remain quiet and hang out in the experience.

A. J. put on the Mind Alive glasses, I set the tint to red and checked the intensity of the light. I suggested that we should adjust the light to be as bright as is comfortable. We then opened the my-Noise app and got the volume just right.

Within the first couple of minutes, A. J. reported that she was in contact with a spirit guide of some kind—one that she had never met before. She proceeded to have an internal conversation with this entity and would periodically verbalize what was going on. She indicated that this guide's name was Damascus* and asked if there was a pink noise setting that could be increased. I adjusted the pink noise to a louder volume until A. J. said, "That's it!"*

After a few more moments of silence, A. J. began dictating the information she was receiving from Damascus. "We tend to think of things as binary, ones and zeros," she said, "on and off, but it is actually three-fold: here, there, and nowhere." She stated that this understanding was the point of the exercises we were doing—to use this understanding to enter a liminal space, a state of consciousness between worlds. A. J. said that the addition of the third noise is a way to train awareness and consciousness.

Initially, A. J. reported that she was focused on the sound of the static as if it were tiny beads falling, but when we added the pink noise, she said she began focusing on the space *between* sounds. She remarked that this was a great tool to use to learn how to focus on

*Note: Toward the end of the session, A. J., asked Damascus about the name, recognizing it as an actual location, the capital of Syria. Damascus indicated that the physical location on Earth is a portal where the veil is thin. The conversation suggested that this spirit guide and the physical location are somehow connected and/or the same.

one part of the sound field while tuning out the others—refining the ability to tune awareness.

A. J. said, "In this reality, time is a constant. Getting people to oscillate out of it is why this sort of thing works."

I asked if the frequency of the sound was important. Still in her meditative state, A. J. asked Damascus and reported that the pink (high-frequency) sounds were helpful for "going outward," while the brown sounds (low frequencies) were helpful for "going inward." The white noise, containing all frequencies, holds the space. I wondered if this was the "here," "there," and "nowhere" that she mentioned earlier?

She said that this technique can help us get out of the third dimension by getting in between two moments. Focusing on the space between sound frequencies, she reported, proved helpful for increasing the ability to hear spiritual messages. She said a similar thing could be done with the glasses by using three different colors. Focusing on the lights could help with seeing spiritual messages.

A. J. then described these exercises as learning how to slow down or speed up time so you can differentiate the specific sounds or colors. She said it is similar to the experience of watching something spinning where it can appear to be standing still or even moving backwards.

During the session debriefing, I asked A. J. if this experience was different in any way from her normal process of connecting to her spirit guides. She indicated that compared to sessions without tech assistance, using the technology made it a lot easier to "tap in." Under normal circumstances, A. J. said, she would have to work hard to get into that kind of connection through meditation or self-hypnosis.

She also reported that connecting to spirit guides is usually an involuntary process for her. In other words, she might be meditating, and the connection just happens without looking for it. With the technology, it felt much more deliberate, as if she could simply set the intention to connect and then effortlessly move into that space.

She suggested that the tech facilitates a "temporary opening" to the spirit world.

While I wasn't sure what I was expecting, it certainly wasn't a full-on conversation with a spirit guide providing guidance on how to use technology to communicate with the spirit world. Of course, there is no way to know for sure if A. J. was communicating with a disembodied entity or if maybe the information was coming from her own subconscious mind. Either way, it was intriguing. At a bare minimum, her experience seems consistent with its use in ghost hunting; that is, using the Ganzfeld approach as a vehicle to communicate with noncorporeal entities.

A. J.'s conversation with Damascus also provided some potential strategies to explore with other people working to develop their ability to listen to the spirit world. Specifically:

1. *Different people may respond to different noise frequencies.* While much of the historical work with Ganzfeld utilized a straightforward, white-noise set of frequencies, it is clear from the exercise with A. J. that adjusting the frequencies may facilitate the process for some individuals. From an intuitive perspective, this makes sense. Not everyone is the same, so we would not expect the exact same procedure to work identically for everyone. Damascus (through A. J.) indicated that the lower frequencies are better for "going inward" whereas the higher frequencies are better for "going outward." This may have implications for specific intentions as well as individual brain wave differences. Later in this chapter I will discuss a few different ways you might experiment with this technique at home.

2. According to Damascus, if you would like to enhance your ability to receive visual information, practice with the lights. If you would like to enhance your ability to hear information, practice with the sounds.

3. The traditional Ganzfeld red-light approach can be modified. Based on the information about sound frequencies, it makes sense that the same idea might be applied to the light. What if the light were white or green instead of red? Would that make any difference? Does it depend on your intention or what your brain likes? What about offering different-colored lights? A. J. suggested three colors, which I do not (yet) have the capability to do, but I can do two colors. Sounds like another good experiment.

4. How you pay attention to the lights and sounds may make a difference as well. A. J. was suggesting shifting attention with both the lights and sounds in such a way as to differentiate between the specific sounds (or colors), focusing on only one while inhibiting the others.

Aperiodic Noise

All of this focus on the sound frequencies reminded me that I had recently attended the Association for Applied Psychophysiology and Biofeedback (AAPB) conference where I stumbled upon a talk on *aperiodic* noise presented by Tiff Thompson, PhD. Dr. Thompson has an impressive resume, including a PhD, two master's degrees, and various high-level certifications in EEG (brain wave) processing and interpretation. Much of her talk was technical, getting into a complicated, mathematics-heavy explanation of what "noise" really is. Of course, given the nature of the conference, the conversation was ultimately concerned with the influence audio noise can have on the brain.

One of the points made by Dr. Thompson was that the brain also contains noise. We tend to focus on the organized, rhythmic signals in the brain that get classified into the brain wave clusters we know so well (such as theta, alpha, gamma, etc.). However, beneath these signals is noise. This is additional activity in the brain that is aperiodic, meaning irregular. It does not have a rhythm and is chaotic. This type of information makes up a significant portion of EEG activity in the brain. When the ratio of signal to noise is high, the brain is organized, effective, and efficient. When the ratio of signal to noise shifts in favor of the noise, this suggests a brain that is less effective at doing its job—it is literally noisy. This is one hypothesis (the neural noise hypothesis) explaining why we see cognitive decline in so many people as they age: the signal to noise ratio has shifted.

Taken on its own, we might look at this information and conclude that it is a bad idea to introduce more noise into the brain. However, there is a phenomenon known as *stochastic resonance*, in which the addition of white noise actually enhances the response of a system. By adding a wide range of frequencies, those frequencies that correspond to the original signal's frequencies will resonate with each other, amplifying the original signal while not amplifying the rest of the white noise. So, the brain is essentially picking out the frequencies that it needs to enhance certain information. This process happens most frequently in systems that have a sensory threshold. When the input is below that threshold, it is not perceived. When energy is added to that input, it can cross the threshold and enter our perceptual awareness.

When applied to the field of parapsychology, this suggests the possibility that the addition of noise may enhance the ability to perceive certain signals. What if there is a "veil," as many esoteric authors suggest, separating our normal conscious awareness from the magical worlds? What if that magical world is always present but is usually just out of reach, just below our awareness?

Learning how to create an opening—a bridge—may be an important aspect to developing and advancing any psi abilities. The addition of noise might help that veil to become just a little bit thinner, making the "signal" from the other side a bit louder and clearer. It's possible that was what A. J. was describing after her session with Damascus. She said the technology made the ability to tap in much easier and more intentional. We could perhaps state this another way: the technology made the information from the other side easier to hear (or see).

A. J. also suggested that certain frequencies were better for this type of work. Perhaps the lower and faster frequencies are more in line with the frequency of psychic information or the "thinning of the veil." Interestingly, when I went back to examine my notes from Dr. Thompson's talk, I found a section indicating that brown noise (lower frequencies) is better to induce sleep and can also stir up suppressed memories from childhood, while pink and white noise both "add energy" to the system. Clearly, different frequencies of noise have different impacts on consciousness. Again, as A. J. put it, slower frequencies seem good for going inward and faster frequencies seem good for going outward. .

So, our initial experiment using a modified version of the Ganzfeld technique appeared to be a success, and there seems to be an interesting correspondence between the information A. J. received and the scientific literature on noise, which increased my confidence that she was tapping into some real knowledge. Of course, at this point we had only tested this on one person. I was curious how other people would respond.

Cammra and Ena

Our next volunteer came to us from an encounter at a local psychic fair. Around the time I decided to write this book, Ray Jackson, my program manager at the NeuroMeditation Institute, and I decided to scope out the local talent. We went to a small psychic fair held in our town and spent a few hours meeting people and attempting to recruit potential volunteers for our explorations. Of all the people we talked to, there was only one who seemed to be a good fit. Cammra Garza had a small table in the back corner of the fair and was offering mediumship readings. Ray spent some time with her and recommended her for our testing. We decided that we would invite Cammra in for a brain map experiment while she was doing both a psychic and mediumship reading for our intern, Jordie Hasha. Cammra also agreed to try a Ganzfeld session to see if that influenced any of her abilities.

Cammra had never met Jordie before and was given no real information about her. After we put the EEG cap on Cammra and collected some baseline data, we got ready for the psychic reading. I asked Jordie not to provide any specific information about herself to Cammra during the reading so that we could feel more confident that any information that came through was not just a good guess or based on something Jordie had disclosed.

Cammra asked if Jordie had any specific intentions for the session and she said she did not. Cammra noted that this sometimes made it a bit more challenging. Apparently, the attitude and intention of the sitter is also an important variable in the equation.

Cammra began: "I am getting that you sing a lot and work with plants. I see that you worked with plant medicines in a previous life. This is an easy connection for you. You do yoga." While these pieces of information may be somewhat generic, you can also imagine there are not many people for whom all three statements are true. Turns out, for Jordie, they were all accurate. While I have never heard Jordie sing, apparently this is something she does at home frequently. As far as plants go, we literally refer to her as our "plant lady." She has brought in thirty plants to our office and takes care of them all. She has over a hundred plants in her home and works part time in the cannabis industry. Not only does Jordie have a regular yoga practice, but she is also a certified yoga instructor. Okay, seems like Cammra's first comments were "hits."

Cammra went on, "I sense that you get physically affected by negative energy in your environment, maybe getting nauseous. You can get overwhelmed by negative energy. I see that you are cautious in relationships." After the session, Jordie confirmed that all of these statements were also true. In fact, she indicated that she often avoids crowds or other situations where she may have contact with people as it makes her feel anxious and uncomfortable.

At this point, Cammra indicated that she was going to switch to mediumship mode, (my term, not hers). After a few minutes, Cammra indicated that she was picking up a woman on Jordie's mother's side of the family. She asked if there was an aunt or someone connected to an aunt that she was close to. Jordie said yes. Cammra went on to describe other details of their relationship, including things they would do together that all seemed to fit.

Afterwards, Jordie was asked if she felt the information was relatively accurate and specific to someone in her life who had died. She said, "Definitely!" While none of this provides any kind of hard evidence, it was impressive that both the psychic and mediumship readings for Jordie appeared to be highly accurate.

A Fast Track to the Other Side?

Next, we moved on to the Ganzfeld experiment. I decided to use a nearly identical setup to the process we established with A. J. The

Ganzfeld glasses were set to red, and the white noise app was filtered to emphasize the lowest and highest frequencies. After a few moments of adjustment, Cammra reported being in a jungle scene that was part of a past-life memory. She commented that the noise pumping through the headphones seemed to enhance her sensory experience. She stated that it was almost as if she could feel the environment with all her senses. Cammra reported that the scene became very misty, and she was transported into a cave.

At this point, Cammra took a deep breath and stated that someone was with her in the cave. It was an older woman named Ena. She had big puffy eyes and messy hair that she wore pulled back. Cammra described that she was unfolding a blanket of some kind and was aware of all of us in the office. After a few moments of silence, Cammra laughed and said Ena was speaking to me and challenging me to come to the cave to meet her.

Cammra provided some additional details of the scene, and I asked her to ask Ena about the purpose of this cave. Cammra internally checked this out and responded that this was an entrance (presumably to another realm or reality). Ena went on to explain that the space she was in exists in the physical world. She stated that there are entrances in nature. Some are near trees; others are near mountains. However, these entrances are not always accessible. They can blink in and out of existence. Ena told us that they can appear with song.

I asked (through Cammra) if Ena would be willing to teach me how to find this opening, to teach me the song. She replied that it is more like a movement. Ena said that it is important to prepare the body to change state and shift into nonordinary reality. Apparently, Ena kept directing her talking points toward me. Cammra seemed surprised by this and suggested that this was somewhat unusual.

Ena communicated that she expects me to use this technology—
the Ganzfeld technique—to connect with her. She stated that this
helps to connect to the doorway quickly, almost like a direct line.
Ena closed by taking Cammra to a large, dark, open body of water
and told us to "remember the water." She stated that this was another
way into the other worlds.

I Remain Curious and Skeptical

This was a lot to process. The curious, open-minded part of me
was excited. Not only did the Ganzfeld experiment seem to work,
but the guide contacted through Cammra wanted to talk to me. Not
only that, but she seemed interested and willing to teach me how
to move into these more subtle worlds. I have always felt that I had
some untapped potential in the psi world, and this sounded like a
direct invitation from the other side.

I still wasn't sure what I was supposed to do. Should I go out in
nature and look for a physical entrance? Was I supposed to use the
Ganzfeld technique? What should I do after I put on the glasses and
the noise? Should I visualize water? Or is the water she talked about
a physical body of water in this reality?

I was skeptical. Even though I have seen a lot of weird things,
the spirit guide work has always been the most challenging for me
to fully believe. This is probably because it is the hardest one to vali-
date scientifically. Sure, we can show that Cammra's (and other mediums') information about a particular person or deceased person
was more accurate than could be expected by chance, yet that doesn't
necessarily mean that they are talking to the spirit of a dead person.
What if they are psychically or telepathically picking up information
from the sender/sitter? Of course, that would also be cool, but it is
not communication with spirits. In situations where people claim to

communicate with spirit guides, or angels, or aliens, or fairies it gets even more difficult. There is currently no way to verify that what the person is reporting has any basis in an actual communication.

It seems to me that one has to make a decision. Either go with it and assume it is a real communication, or dismiss it as imagination or subconscious aspects of the self. In this case, I decided to go with it. It certainly couldn't hurt anything, and what if I could, somehow, connect with Ena (or some other guide) through this process?

Stepping back and looking at Cammra's process with a wider lens, it struck me that there were some similarities and consistencies with A. J.'s experience. First, they both made contact with some kind of guide/teacher. In both cases, the guide/teacher was providing information for me or to me. Finally, both A. J. and Cammra (and their guides) described the Ganzfeld technique as a fast track to connecting to the other side.

Jeff and Ti

Obviously, I wanted to try this myself. I have never considered myself particularly gifted at communicating with the spirit world, although there have been rare instances where I felt or saw such a connection during certain states of meditation. Maybe Ganzfeld would help me access this state more easily. And so I devised an experiment for myself. It began with emphasizing the lower frequencies of the noise app and splitting the color of the Ganzfeld into red on one visual field and green on the other. I was thinking about what I heard from A. J./Damascus as well as Cammra/Ena. Maybe having more than one color in the glasses would assist in "seeing between the colors," and maybe focusing on the deeper, ocean-sounding aspects of the noise would somehow connect me to the water aspect that Ena suggested? It was worth a shot.

After about ten minutes, I wasn't noticing much and felt like these weren't the best settings. I shifted the glasses to full red and cranked the intensity to 100 percent. I also added more of the high frequencies in the noise app. Very quickly, I noticed a shift in how I felt. It was easier to quiet my analytic mind, and I sensed an automatic shift in my awareness.

I (internally) asked if any guides were around. I asked for Ena to assist me. I asked for any kind of communication, and then I just sat. Aside from a slightly altered state of consciousness, there weren't any fireworks. I wasn't whisked away into another realm, and I didn't hear a voice from the angels come down and tell me what to do. However, I did get a flash of a face, and when I asked who was there I heard, in my own voice, "Ti."

Not much else happened, and I stopped the session after another ten minutes or so. I was disappointed. Why didn't this work for me when it seemed to work so well for virtually everyone else I had tried it with?

Later as I was driving to work, I was aware that I somehow felt different. I felt more aware of the subtleties around me in my environment. I had a sense that there was an energy just out of perception that was just enough to be tangible—if I paid close attention to it. The fact that I wasn't looking for this feeling and had been feeling pretty discouraged gave me a bit more confidence that my subtle energy perception was real and not just something I was making up due to wishful thinking. Perhaps the glasses did start to loosen things up. Maybe I didn't spend enough time in the session, or maybe there are times when the impact of the Ganzfeld is delayed. Rather than happening during the session, perhaps it has its psychic-enhancing effects later in the day.

Into the Mystic

A few months after my self-experiment, I started working with psychic medium Jeannine Kim. Along with two other psychics-in-training, we would meet on Zoom to explore our abilities, receive feedback from Jeannine, and discuss ways that we might incorporate some of my research with technology. To this end, everyone in the group purchased a set of AVE glasses so we could experiment with the Ganzfeld settings. We decided to try a kind of remote viewing/psychic exercise. Jeannine picked twelve famous people that all of us would recognize. She did not tell us who any of these people were, but wrote their names on slips of paper, and then placed the slips inside individual, opaque envelopes that were then labeled with a number from one to twelve. Jeannine assigned each of us a number/person to see if we could pick up any information related to that individual. As a group, we also chose a couple of numbers that all of us would tune in to. Each week, our homework was to spend time tuning in to the assigned numbers and writing down our impressions. Sometimes we would use the Ganzfeld glasses and sometimes we would not. This would help clarify if the addition of light and noise seemed to help. When we met the next time, we would compare notes and then Jeannine would reveal the name behind the number. She also gave us some tips when doing this type of psychic work.

1. Set an intention before each reading for the highest and greatest frequency and the clearest message (for the individual you are reading in the envelope).

2. Set an intention as a witness and messenger to bring their news forward. Intend to remain open, gently curious, committed, consistent, and empty.

3. Clear your energy field when you begin and especially when you finish, letting go of the energy connected to the individual you are reading.*

4. Remember that it is just energy that you are tuning in to. The more attuned you are to your energy in every moment, the more you will be attuned to the energy at hand. Understand that you read differently than others and that is beautiful; let it be that way.

5. Let yourself feel silly. Just go for it. Color outside of the lines and let the process unfold naturally and in whatever way it happens.

6. Empty your mind and trust what you see.

The Trials

The first few times I attempted to tune in without the glasses. I got nothing. If I was trying for a quiet mind meditation, I would have been very successful. I didn't see anything; I didn't get any impressions or feelings. I had followed Jeannine's suggestions to the best of my ability but felt like maybe I wasn't directing my attention in the right way. I switched to using the Ganzfeld glasses to see if this would make any difference. I figured it was worth a try.

Envelope Eight

For my first Ganzfeld session with this psychic envelope experiment, I decided to focus my attention on the person identified in envelope eight. I initially saw an image of Jeannine. I figured that had nothing to do with the name in the envelope but was instead my mind making associations and connections related to the exercise.

*Jeannine Kim provides detailed instructions about this cleansing meditation on her website: www.jeanninekim.com/sacredclearing.

In fact, this awareness was helpful as it assisted me in distinguishing between "thoughts" and "information." Thoughts were a result of my brain thinking, analyzing, and trying to make sense out of the experience. Information, on the other hand, was data that the conscious mind did not construct. It seemed to come from nowhere. I often recognize information like this when I see or feel things that seem to have nothing to do with what I am thinking and seem somewhat random. At any rate, the next image I saw was a cabin in the woods. I had the feeling of someone being alone. I saw a dark-haired man. I felt he lived a simple life. He was not married and did not have any kids. I thought of the character Ron Swanson from the TV show *Parks and Rec*. That was it. I didn't get anything else, and it was difficult for me to tell how much of the information was thoughts versus information. Certainly, the initial impression of the cabin felt like information, but then it seemed like my mind got involved and started trying to make connections. Regardless, it was interesting that I seemed to be getting more images and impressions in this session compared to the sessions without the glasses.

The name inside the envelope was Mother Teresa. On the surface, doesn't look like I had much of a match, although it was pointed out that I was picking up on a feeling of someone living a simple life, with no kids, and being alone. This might fit. Of course, there was a lot of other information that did not necessarily match. Cabin in the woods? Male figure? Was this just my mind making up stuff and finding connections where none existed? I don't know, but I did feel like I was beginning to learn how this process worked. It was not straightforward or linear. It made sense that the more creative and unrestricted aspects of the mind are likely to communicate in symbols and feelings rather than presenting a direct, obvious connection (at least for a beginner like me). It seemed a bit like the way dreams work. Sometimes they are clear and obvious, but usually they are disguised. The theme and the tone are what are more important than the exact content. If this was true, it was going to make the process of interpreting received information extremely difficult.

Envelope Five

The next session I started with my typical morning meditation routine. I did some stretching, qigong energy clearing, an invocation, brief meditation, and then put on the Ganzfeld glasses and attempted to tune in to envelope number five. Again, with the Ganzfeld glasses it was much easier for me to receive images. I got the impression of a dark-complected man. He had a beard and moustache. I thought maybe he was Arab. His face reminded me of a friend. Maybe he wasn't Arab, maybe he was Latin. I didn't have enough clarity to know for sure. Next, I had the image of a sea of treetops with a bird-like thing over the top that was moving in a strange way, diagonally shifting back and forth. Not the flight of a normal bird, maybe a drone or something? This shifted to a black-and-white impression that turned into a skeleton-type mask image. Next, I received a fleeting image of an angry dog/wolf, maybe rabid. This shifted into a feeling of moving through a tunnel. My next thought was: *Is this someone who is dead?*

I remember at the time feeling as if this was a lot of random impressions. They didn't seem to go together in any obvious way or tell any kind of clear narrative. I attempted to follow Jeannine's advice and just go with any impressions I received without judgment. When our group met again and we discussed the envelopes and our practices, it was revealed that the name in envelope five was Princess Diana. My initial feeling was that I had completely missed the mark. Nothing I saw or felt seemed to even be remotely related.

After we all shared, Jeannine pointed out that Princess Diana was killed in a tunnel and was with her lover, Dodi al-Fayed, an Egyptian billionaire. Okay, that was interesting. Not sure how the other details may fit. Were the rabid dogs symbolic of the paparazzi following her car just before the accident? Was the bird-like drone representative of helicopters flying over the scene? I have no idea. Maybe I was tapping into something and maybe it was a coincidence, and I was looking for meaning and making the details fit. This whole process reinforced the idea that the way information comes through is not necessarily straightforward. How do you learn to trust these images and impressions when they may be only loosely connected? The idea that maybe I was getting some "hits" on Princess Diana was compelling. I wanted to believe that I was receiving information, but it was hard to trust. I could feel the different parts of myself competing: the scientific part of me writing this off as wishful thinking, while the "believer" part saw this as confirmation that I could develop these skills and that the Ganzfeld seemed to help. I knew I would not have any clear answers and attempted to let the experience go. I would file it away and see what happened next. Turns out, I didn't have to wait long.

Synchronicity

The next day, while working on an unrelated chapter of this book, I was trying to find some information about meditation and psychic abilities. I had a vague recollection that Dean Radin talked about this in his book *Entangled Minds*. I hadn't looked at this book in years but found my copy on the bookshelf and began leafing through it. I noticed that there were ten or so pages that had been dog-eared from my previous reading of the book. I checked each of these pages, assuming that I had already tagged something useful. As I flipped through, scanning for the word *meditation*, I caught a subheading midway through the book: Princess Diana. While this was obviously not the reason I tagged this page, it seemed like a strong coincidence. Of all the books I own, this is likely to be the only reference to Princess Diana and her death. It also mentioned her companion Dodi al-Fayed. After my initial shock, I went back to read that section. In the third paragraph it mentions that "by un-happy coincidence, Mother Teresa died a few days after Princess Diana." (And of course, during her short lifetime, Princess Diana did meet with Mother Teresa.)

This got me thinking about other experiences I have had in the past during meditation, in altered states of consciousness, and with technology-based approaches. Typically, I don't see things. That is not my strong suit. I also don't tend to hear things from another voice. I do, however, tend to hear things in my own voice and often have a strong sense of "knowing." (As clairvoyance means clear seeing, the term for clear knowing is *claircognizance*.) For me, this is more like a feeling or a certainty—about some idea or thought or awareness—that seems to suddenly jump into my mind. Some might call this intuition, but it seems to have a different quality to it than simply having a hunch about something. At times, these experiences are so intense that the information received feels absolutely true, to the point where I feel compelled to act on these feelings/knowings. So far, every time I have listened, I have been rewarded.

Case in point: several years ago, after a significant altered states experience, I had a clear feeling that I was supposed to leave my position at the University of Missouri. This would mean leaving a comfortable job that I had spent a year creating. My position as a health psychologist allowed me to teach biofeedback, meditation, and qigong while receiving a good salary and excellent benefits. What would I do if I left? I had no backup plan. I didn't have something else lined up and no perspective on what I was supposed to do instead. Yet the feeling was so intense and clear that I felt I needed to honor this information.

After a bit of internal wrestling, I left my comfy position and returned to private practice. Within a year I had published my first book, *Meditation Interventions to Rewire the Brain,* was teaching workshops around the country, and forming the NeuroMeditation Institute, none of which could have happened if I had stayed at the university.

When I look back at my experience, I have to wonder about my own expectations. I was assuming I would receive information the same way that A. J. and Cammra do—by seeing or hearing—but this is not the way I tend to operate. This doesn't necessarily mean that I can't develop those other abilities, and in fact when I practice certain kinds of psi skills I do see or hear things, but my experience does suggest that I am much more likely to connect or receive information through this *felt* sense rather than through what I judge to be a more interesting and "sexy" mode of communication. I often forget that clear knowing, or claircognizance, is one of the major modes of receiving psi information.

The Brain on Ganzfeld

As part of the process for both Cammra and A. J., I had collected EEG data before and during the Ganzfeld experiment. Not only did I want to see if there were any significant changes in brain wave activity, but also if there were any similarities between the two case studies. The answers are "yes" and "yes."

When we compare Cammra's baseline EEG recording to her recording during the Ganzfeld experiment, we see a significant increase of both slow (delta and theta) and fast (high beta) brain wave activity. Essentially, the areas that were already slightly elevated at baseline became exaggerated with the addition of the lights and sound.

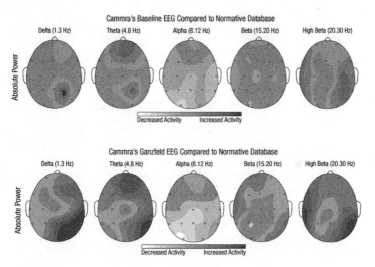

For each EEG band (e.g., delta, theta, etc.), the heat maps above show how much electrical brain activity Cammra has compared to a normative database. Darker shades indicate that there is more activity compared to average; lighter shades indicate less activity compared to average. A quick glance at these two maps shows a significant increase of delta and theta in the right rear quadrant with the addition of the Ganzfeld process. There is also an increase of theta in frontal regions and an increase of high beta in the right rear quadrant as well as the left temporal areas (by the ear).

The fact that the same basic patterns exist in both recordings may suggest that Cammra's natural state was enhanced during the Ganzfeld session. Although I am speculating, this makes sense as Cammra privately indicated that she often must actively work to "turn off" her medium and psychic skills, suggesting that her brain is primed for this kind of activity, as if it is waiting for an opportunity to turn on and tune in. Cammra also indicated that the Ganzfeld session seemed like a fast track to that state. Perhaps the additional energy from the light and sound simply enhances the already-present tendencies.

Interestingly, we saw a similar process with A. J. At baseline, she showed increased levels of delta in several regions (including the right rear quadrant), increased global theta, with a bit more in the frontal regions, and increased beta and high beta. Different from Cammra, A. J.'s fast activity was much more focused in frontal regions and leaning to the left.

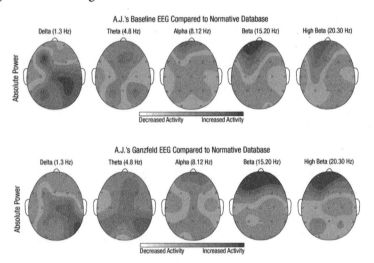

Like Cammra, during the Ganzfeld session, the slow wave activity seemed to become focused in the right rear quadrant and the fast activity became exaggerated. So, both of our case studies showed similar patterns with slow activity. Fast activity increased for both participants but showed up in different areas. For Cammra it appeared in the right rear quadrant, the same as the slow activity, whereas with A. J. it showed up in prefrontal/frontal areas.

What can we discern from the specific brain waves and the locations involved? Clearly, there is an emphasis on the right rear region in all of the brain wave bands mentioned. This area of the brain is often referred to as the temporoparietal junction or the right parietal lobe and has shown up in other cases discussed in this book. As a reminder, this part of the brain seems to be important for defining boundaries between self and other.

I have hypothesized that this part of the brain "goes offline" or in some other way seemingly "misbehaves" (or behaves differently) during many forms of spirit communication. In fact, you may recall that I highlighted this area when discussing Janet Mayer (Chapter 1). Essentially, shifting the normal functioning of this part of the brain appears to open a connection between self and other, whether this is a guide, angel, or disembodied spirit, essentially allowing the medium to connect with energies outside of the normal little ego self.

The fact that both slow and fast brain waves increased suggests both shutting down and activating processes. How is this possible? Maybe the normal function is shutting down, while other aspects involved with connecting outside of oneself are activated. Of note here, remember that the noise we used was emphasizing the slowest and the fastest frequencies simultaneously.

We also see an increase of theta activity in frontal regions. This is most obvious in Cammra's brain wave comparisons but occurred with A. J.'s as well. Frontal theta is often associated with "spacing out," and ADHD-type symptoms. Essentially, when there is an excess amount of frontal theta it is often a sign that those areas are not working as effectively or as efficiently as we would like. This increase of frontal theta activity could be viewed as the frontal lobes shutting down. This is important as this part of the brain is largely responsible for high-level cognition as well as inhibiting information.

The reduction of frontal lobe activity is another common finding in the world of psi. For A. J., she simultaneously had this frontal theta (shutting down) while also showing increased fast activity (activating).

What to Make of the Ganzfeld Technique

From a practical standpoint, it seems safe to say that the use of the Ganzfeld technique has the potential to facilitate spirit communication. These effects seem to be the strongest for people who already have a history of spirit communication. For these folks, connection to spirit is consistently enhanced by the use of the Ganzfeld technique. For those who do not claim any talent related to spirit communication, the results are inconsistent. In some cases, people report intense experiences and contact with various entities, and in other cases it is simply a relaxing meditation experience.

It makes sense that self-identified mediums would find this approach helpful as it may be enhancing brain tendencies that are already present. In the case studies presented in this chapter, both Cammra and A. J. showed increased activation of delta, theta, and high beta activity in the right rear quadrant of the brain. These areas were already highly engaged before the Ganzfeld technique and are related to boundaries between self and other. Based on other case studies I have conducted, this region of the brain appears to be important for this particular set of skills.

This finding may also help explain why nonmediums have varied results. If their baseline brain patterns are not consistent with spirit communication, adding energy to the system won't necessarily catapult them to the "other side." In this regard, it may be necessary for nonmediums to do additional training to shift the activity patterns in that right rear quadrant of the brain. This might involve using additional technology, such as *transcranial direct current stimulation* (tDCS), or low-power electromagnetic stimulation, which will be discussed in other chapters.

There also seems to be some preliminary evidence that different people respond to different "colors" of noise. While the traditional Ganzfeld approach uses white noise and seems to work, my experience has been that certain individuals find it easier to tap in when the frequencies are shifted. In general, an emphasis on the lower and higher frequencies at the same time seems to show the best results. However, there is a fair amount of individual variability in preferred settings for noise. This may be something that each person needs to experiment with for themselves, noting their experience with different noise combinations, and tuning into a felt sense of which frequency combinations feel right.

Finally, it seems important to recognize that there are (at least) six different ways of receiving psi information:

1. Clear seeing (clairvoyance)

2. Clear hearing (clairaudience)

3. Clear feeling (clairsentience)

4. Clear knowing (claircognizance).

5. Clear tasting (clairgustance)

6. Clear smelling (clairsalience)

In general, it appears that most practitioners of psi have a preferred modality, a way that they most easily receive information. However, with focus and practice, psychics and mediums can learn to enhance their other "clair" senses as well.

From the presented cases, it is clear that A. J. tends to hear information. In fact, her intention for the Ganzfeld exercise was to better develop her visualization skills. On the other hand, Cammra seems to work with both clairvoyance and clairaudience. I receive information more through knowing and feeling, which is clairsentience. It makes good sense that different people would receive information through different sensory modalities.

Perhaps the brain patterns associated with psi *require* certain parts of the brain to be more (or less) engaged (like the right rear quadrant and the frontal lobes), while the specific sensory modality used is a secondary process. The necessary brain changes relate to *what* is happening, while the sensory brain changes relate to *how* it is happening. This appears to be how meditation works. For example, if you are doing a focus style of meditation, you can focus on a visual image (a symbol such as a mandala, for example), a sound (mantra), a sensation (belly expanding and contracting), or a thought (prayer). All of those require different sensory systems and consequently different brain regions. However, *all* of them engage the same brain region related to sustaining attention, and *all* of them disengage the same brain region related to thinking about the self.

Perhaps part of the trick with developing psi skills is identifying which modality is your strongest and leaning into it.

Try It Yourself

If you are interested in trying the Ganzfeld technique, you will need a way to create the red glow effect as well as a source of white noise.

Creating the Ganzfeld Visual Field

To create a red homogenous visual field, you will need a source of red light as well as a semi-translucent eye-covering or specialized pair of glasses.

Ping-Pong Balls. For a simple, cost-efficient approach, you could follow the original instructions and cut a ping-pong ball in half, taping each half over one of your eyes. It is important for the full effect to make sure that there are no gaps around the edges that might allow in extra light. Once the ping-pong ball halves are secure, you face a red light source, which could be a red lightbulb in a lamp.

Goggles. A slightly more advanced approach would involve purchasing safety goggles or swim goggles and spray painting them with a frosted-glass style of spray paint. I have also seen blogs where people describe simply attaching a piece of white paper around goggles, which also seems to work.

Audio-Visual Entrainment Glasses. For a more sophisticated approach, I recommend purchasing a Spectrum eyeset, which is available from our website. These glasses have lights built directly into the eyeset and are capable of many different types of light and sound stimulation that can help with mood, sleep, or energy. The glasses plug into a USB port on a computer and are activated by a simple software program which can be downloaded from the Microsoft app store (DAVID Live). The glasses can be purchased at: www.psychicmindscience.com.

Ganzfeld Setting in David Live

In the uppermost section of the software interface, you simply slide the frequency bar all the way to the right side until it says "Ganzfeld"; this will cause the lights to stay on at a constant rate. In the third section of the interface, you can choose the color of lights for the left and right visual fields.

White Noise

Historically, the noise used in Ganzfeld experiments is white noise, which, you may recall, is static from all of the possible auditory frequencies. Sources of white noise are easy to find. In fact, you can do a YouTube search for the term and discover dozens of options. You can also find numerous apps for your smartphone, many of which are free. Any of these programs should work fine. The larger consideration is that you have headphones that will block out external sounds. Any auditory intrusion, such as your roommates, music, or cars driving by, is likely to disrupt your experience.

Ganzfeld Tips and Modifications

Relax. It is important to make yourself as comfortable as possible before your session. If the body is not relaxed, the mind will not relax. Many people find it helpful to lie down or sit in a recliner chair. Kick your shoes off, breathe slowly and deeply, put a slight grin on your face, and then begin the session.

Take Notes. The state of consciousness in a Ganzfeld session often feels a bit dreamy. Because of this, some people have a hard time recalling what happened during their session, or the information fades quickly. I recommend setting up your phone to record throughout the session. This will allow you to stay engaged in the session and verbally report anything you see, hear, or experience. The recording can be played back later to jog your memory. If this setup doesn't work for you, you might try keeping a journal next to your meditation station and making notes as soon as the session ends.

Length of Session. While most people begin experiencing some kind of visual/auditory alteration after five to seven minutes, we recommend sessions between fifteen and twenty-five minutes to get the full effect.

Intensity. I usually recommend adjusting the brightness of the light and volume of the noise as high as is comfortable. In general, a certain level of stimulation is necessary to create the effect. However, we all have different thresholds for sensory input. If either the lights or the noise becomes too intense, leading to agitation or anxiety, you can always discontinue the session or reduce the intensity/volume to a level that is comfortable.

Experiment with Color. While this technique is usually done with red light, it can be fun to try something different. In our experiments at the NeuroMeditation Institute, we have found that white light also works well. If you are using the Spectrum eyeset, you can choose any number of colors and even modify the settings so that one color is going to the right hemisphere of the brain, while another color is going to the left. It might be interesting to use a stimulating color for the right hemisphere (left visual field), such as red, and a cooling color for the left hemisphere (right visual field), such as blue.

Experiment with Noise. While this technique has traditionally been done using white noise, it can also be useful to experiment with pink or brown noise. These represent the higher and lower frequency ranges, respectively. Similar to white noise, you can find apps or YouTube videos that provide this type of stimulation. If you are interested in a deeper exploration of the effects of different noise frequencies, I recommend purchasing an app such as myNoise, which includes dozens of noise generators. One of these is a colored noise generator, which allows you to modify each of ten different frequency ranges within the white-noise spectrum. With this app you can include both brown and pink noise or any other combination that may be of interest.

Ground Yourself Afterward. Because these sessions can be somewhat disorienting, it is a good idea to allow some time after a session to reconnect fully with your body. You can encourage this by breathing a bit more intentionally and deeply, moving the fingers and toes, arms, and legs, sharing your experience with someone else, and engaging your senses (what do you see, feel, smell, hear), preferably outside in nature. We also recommend drinking water and eating something with "grounding" energy, such as nuts (or chocolate).

A Note of Caution

While the cases we described in this chapter were mostly pleasant and benign, it is not uncommon for someone to have an experience that is somewhat unsettling or even frightening. Some people report a "creepy" feeling or a sense of being watched. Other people become overwhelmed by the sensory input and feel anxious. In general, if you take the glasses off and stop the session, these effects quickly dissipate. However, because this technique and its effects disrupt normal sensory processing, we do not recommend trying this technique if you have any serious mental health concerns, such as bipolar disorder or schizophrenia. While there is no evidence that the strobing light from these devices can cause seizures, if you have a pre-existing seizure disorder, it is probably best to avoid this technique. In addition—similar to a Ouija board—it is also possible that this technique may open connections to other entities that were not specifically invited. While we have only witnessed one situation that may fall in this category, it seems important to mention. See below.

Ray and the Wet Woman

Shortly after we began running a variety of Ganzfeld experiments in our office, my program manager, Ray, stayed late one day to run a session on herself. When I saw her the next day she said, "I have to tell you about my session, it was totally freaky and felt a little bit like a horror movie." Here is what happened in her words:

I began the experiment with red light in both eyes, set on the brightest setting. At first, I simply sat and wondered how long it would take to "snap in." Then I became aware that my eyes were moving in a REM pattern under the glasses. I realized then that I must not be fully conscious. It was slightly agitating, and my heart began to beat more rapidly. Suddenly, I had a vision of falling backward into water and watching the bubbles surround the body of a slender woman with long, straight, brown hair, and a dark blue dress with small flowers on it. Next, I got an image of myself looking into my bathroom mirror, and my own reflection moving incongruently with myself. She stepped aside to show me the bathtub. There was the woman who had fallen into the water standing behind me in the tub. She was wet. Water dripped from her fingers.

Next, I was back in my childhood home. I was in the bathtub, a memory from seven or eight years old, and I was staring into the water of the tub. I begin to see another world appear in the water, a large building with yellow-beige bricks. It wasn't a pyramid but had a large base and square bricks that came up like a tall rectangle sitting on top of a square. The sky was faded yellow. This is something that I do remember happening in real life. However, next in my vision, I was pulled backward by the hand of "the Wet Woman" into the tub and "child me" repeats the same backward splashing. Now I am in a dream world that I have seen in my childhood. I feel very warm.

All of my vision comes in black and white, as if watching an old film, but there is a blood-like red film that looks as if it has been poured over the projector reel in my mind. I have had this dream before, too. I am back inside of my parents' room, but I have never seen this part of their room before. Young me is chanting. I am looking down at her from someone else's eyes. The leaves on the monkey grass outside rustle. The monkey grass along the path always leads me to a deeper dream world. I become aware of this. The Wet Woman is walking in my old apartment when I first moved to Eugene. I awake to her there, in my vision. I use my intention to pull wet grass from her hair. *She just needs love and someone to listen,* I think to myself. I see her eyes, bright blue, bloodshot. She keeps staring into me. My mind tells her: *I made you. I will not be afraid.* For a moment she has large, innocent blue eyes. But she turns back into the Wet Woman. I see her walking up the stairs to my office. Now she is in the hall. She steps into the lobby, and she is looking through the window at me. I tell her again that I made her. I tell her that she must be an angel. She looks young and beautiful. So much light bursts out of my body like a starburst. It has no feeling. I do not feel pleasant, though I suspect that I should. I do not feel exhaustion or fear. She has become the Wet Woman again. Her consciousness moves through the window and her wild eyes try to get close to mine. I feel her on the other side of my office door. I pull off the glasses. This vision is over.

It is a little hard to tell from her description of the experience, but it was clear from the way she told the story that she felt a bit like she was being haunted, like this "Wet Woman" kept following her into different parts of her life and had some kind of malicious intent. This was the first time I had heard of something like this occurring in a Ganzfeld session, and I just assumed it was kind of a dreamlike vision, a storyline made up by the mind. However, two weeks after this incident, we had a follow-up session with Cammra. She had barely sat down before turning to Ray and saying, "Do you have something attached to you?" I had no idea what she was talking about, but Ray nodded her head affirmatively. Cammra indicated that there was an energy that was attached to Ray, and it was making it hard for her to concentrate. She asked Ray to leave the room and commented that the energy left the room with her. Cammra suggested that we needed to do some kind of energetic clearing of the space, which we did by ringing bells and burning palo santo. (This is a South American tree that produces a fragrant resin when burned. It is often used to clear out negative energy.) I still had no idea what was going on until checking in with Ray a little later.

When I asked Ray what this was all about, she indicated that she has continued to "see" and feel the presence of the Wet Woman in our office since that earlier Ganzfeld session. I wondered why she hadn't mentioned this. When I asked her about it later, she told me that it was hard to trust her perceptions. Maybe she was just remembering scenes from her Ganzfeld session or imagining things. She also didn't want to sound crazy. Obviously, I am open to all sorts of strange happenings, but I understood her trepidation. It can be a little much to announce that you have a negative energy attached to you that is folloing you around at the office.

We also had two strange incidents with water in our office around the same time as this meeting. In a separate room that we use for meditation groups and altered-state experiences, we have a water fountain and a variety of plants. Jordie our intern and plant care-taker, was doing her weekly watering routine when she discovered that the fountain was leaking and came to tell me. I followed her into the room and saw that there was water all over the platform. This system had been in place for many months with no other incidents like this. We cleaned it up and watched water seep out of the bottom of the fountain, forming a new pool on the table. As the water was all contained in the base of the fountain, I wasn't sure what we could do besides turn it off, drain it, and look for cracks. We turned it off, but then ran out of time for the next steps. I returned the next day and there was no mess. I turned the fountain back on and everything worked fine, and we have had no additional issues since then.

Two days later, I left my office to meet someone in our waiting room. I was wearing only socks (no shoes), and as I stepped into the main area, my feet became soaked. I felt around on the floor and located a large wet area of the carpet, apparently coming from our watercooler. I opened the watercooler cabinet and found the five-gallon water bottle empty. My office manager told me she had just replaced this in the last couple days, suggesting that the entire thing had leaked onto the floor. We replaced the bottle and tested the empty one, finding a small leak in the bottom of the bottle. This seemed to be the end of it. After this, there were no additional in-cidents and Ray indicated that she no longer felt the Wet Woman hanging around. As a scientist, I'm not sure what to make of this. I know correlation is not causality; just because odd things happen in proximity to each other does not necessarily mean they are related. On the other hand, this type of experience is not uncommon in the world of ghost hunting.

I shared this story with several of my psychic/medium friends, asking if this kind of thing ever happens to them. Surely, this kind of experience is commonplace if you are opening to the spirit world on a daily basis. Interestingly, all of them acknowledged that this kind of thing does happen, but they all denied that it happens to them. That seemed curious. When I inquired a bit more on the subject, it became clear that everyone I talked with was very clear in their intentions for their session, calling in their spirit guides and guardians, and surrounding themselves with white light as a form of spiritual/energetic protection before going into a reading or session. Based on their practice and advice, I would suggest developing a protection ritual, similar to the one below, and using it prior to any of this work.

White Light Protection Ritual

1. Close your eyes and imagine a pure white light emanating from above, from the center of the Universe. Imagine this white light as pure positive energy.

2. Breath slowly and deeply, visualizing this white light dropping down through the crown of the head and filling the entire body with bright, healing energy.

3. As the body fills with this light, imagine it overflowing, and covering the outside of the body, growing into a bubble of protective energy that completely surrounds you on all sides, above, and below.

4. Either at the end of the light visualization or throughout steps 1 to 3, speak your intent several times (e.g. "I am completely cleansed, healed, and protected").

CHAPTER FOUR

GAMMA WAVES AND ESP

THE SECOND NIGHT at the Forever Family Conference, where I presented with Janet Mayer, I had several dreams. This is significant because I rarely remember my dreams. This one was vivid and distinct. It was one of those dreams that, when it happens, makes you say to yourself, *I need to remember this*. Dreams like this have a feeling, like something is happening beyond a random firing of neurons or memory consolidation.

In this dream, I found a bird that had been injured and was unable to fly. Toward the end of the dream, I held the bird in both hands and made up-and-down movements with my arms, attempting to give the bird the feeling of its wings moving up and down, catching the air. I did this for some time and then at the natural place in the movement cycle, I let the bird go. The bird did not fly. It landed on the ground, but not hard. It was like it was able to gain some friction with the air, at least enough to minimize the impact of hitting the ground. I wasn't sure what, if anything, this meant, but it stuck with me all morning.

After breakfast, I met Dr. Diane Hennacy Powell. She is a psychiatrist, trained at Johns Hopkins University, and was previously on the faculty at Harvard. She was one of the guest speakers at the conference and author of the book *The ESP Enigma*. We ended up walking together, back to the auditorium where the conference was being held. During the walk we stopped at a water feature to watch a hummingbird bathing itself. Dr. Powell proceeded to tell me a story about a time that a young raven had fallen out of its nest onto her front walk. She took it in and built an area for it and provided toys for entertainment. Her intention was to release it back into her yard since that was where it was born. Once the raven was fully fledged, she tried to help it fly, but it was not able. At this point she contacted a bird sanctuary to give it a permanent home and additional care.

As she was telling this story, I was a little freaked out. The timing of this was very conspicuous. Her story and my dream were extraordinarily similar. Could it have been a coincidence? I suppose. Were the stories identical? No. But it felt meaningful, as in "full of meaning." So, what was the meaning? What is the point? Was I supposed to do something with Diane? Was this a signal that we were connected in some way? Was this just a clue that such things are possible, or maybe that I am capable of having psychic experiences? I have no idea, but, interestingly, after the conference, Diane and I began discussions about collaborative research which manifested nearly eight years later.

Premonitions

Fast forward one year. I was returning home from a trip to New York where I was working with four different mediums over two days. I flew into St. Louis where I was supposed to catch a shuttle back to Columbia, Missouri, which is nearly two hours away.

Unfortunately, my flight home was delayed, which caused me to miss my shuttle, which happened to be the last shuttle of the night. The shuttle company was closed, and I knew I would not be able to get another one until the next morning. Fortunately, my dad lives in St. Louis, and he graciously came and rescued me from a long night at the airport. Before going to bed at his house, we made plans for the next morning. Knowing that I would wake much earlier than he would, he gave me the keys to his car so I could go to the local coffee shop, have breakfast, and make arrangements for my trip home.

The next morning, I left during the morning rush hour for the two-mile trip to Panera. As I turned onto the main road, the woman in front of me began to pull out and then stopped. I was looking to my left for any oncoming traffic, didn't see her stop in time, and smashed right into her back end. No one was hurt, but I was frustrated and embarrassed. After exchanging insurance info and getting my coffee and bagel, I returned to Dad's house to deliver the news that I had wrecked his car. After the initial surprise wore off, he said, "I had a premonition about that." When I inquired further, he described a dream he had sometime during the night or early morning that involved me, his car, and the car being wrecked. He seemed a bit surprised when he realized this and commented that he could have had the dream/premonition at the same time as the accident.

This was interesting because my dad is a super-skeptic. In fact, every time I talk to him about this branch of my work, he acknowledges the possibility that some people may be able to access knowledge from an unknown source but is also clear that he generally doesn't trust people who claim to be psychics or mediums. I joked that he was psychic, which prompted him to tell another story. Many years earlier, he woke up suddenly one morning and knew that my sister (who lived in another state at the time) was going to show up that day for a surprise visit. As far as he could remember, there had not been any clues that this might be a possibility. Not only did the dream seem random, but the feeling behind it was so strong, that Dad got up and started cleaning the house, convinced that Nikki would show up unannounced—which she did later that day.

This kind of experience is common and often occurs for people at the time of a crisis. In fact, the founders of Forever Family Foundation, Bob and Phran Ginsberg, had one of these experiences the day their daughter was killed in a car accident. Bob told me that the morning of the accident, Phran woke up in a panic. Bob helped calm her down and asked what was wrong. She said that something bad was going to happen today but had no specific details. Even though Bob was generally skeptical of information that came in this manner, he had already witnessed Phran's ability to "know" things on several other occasions. Throughout the day, Bob checked on all three of his kids several times to make sure they were okay. The day passed without incident and the entire family went out for dinner. After dinner, their daughter, Bailey, rode home with her brother. Bob and Phran were following them back to the house and came upon the accident. Both kids were severely injured and rushed to separate hospitals. Their son was in a coma and suffered significant head injuries but eventually recovered. Their daughter died two hours after the accident.

Synchronicity

This ability to see and feel into a future event is more common than we might imagine. In fact, a YouGov survey of 1,189 US adults found that over one-third of the respondents reported experiencing a premonition or accurate prediction of the future (2017). Clearly, this isn't an ability that is limited to a select few. What is going on here? How can we tune in to feelings or events that we should not know about—events that have not yet happened? Carl Jung termed such experiences *synchronicities*, which he defined as a meaningful coincidence of outer and inner events that are not causally connected (2010). In the cases above, the dreams (inner event) were connected to specific events that happened later the same day (outer event). The reason that such connections are possible may be related to the concept of quantum entanglement. Essentially, this theory (applied to psychic phenomena) suggests that the mind is not just a product of neural firing but is actually connected to all information throughout space and time. Under the right conditions, the mind can access bits of this information and translate it into an internal thought, image, or feeling. If this theory is accurate, it would explain not only how these abilities are possible, but also why some people seem more tuned in than others. If the brain is essentially a receiver of information from the Universe, it makes sense that some brains may be more naturally adept at accessing or translating this information. It also makes sense that we may be more open to this type of information during altered states of consciousness, such as deep sleep. The information is there, we simply must figure out how to connect and listen.

While psychic events like those described above are often spontaneous, occurring during sleep, I was interested in exploring the conscious ability to tap into this universal field of information. In addition, I wanted to approach this from a more scientific perspective. As compelling as the dream-premonitions stories are, there are no current methods to demonstrate that these are statistically meaningful or "real" beyond mere coincidence.

Testing ESP (Extrasensory Perception)

The easiest and most well-researched approach to testing psychic skills involves the use of a specialized deck of cards named after their creator, perceptual psychologist Karl Zener, PhD. The Zener deck is made up of twenty-five cards. Each card contains one of five possible symbols, including a circle, square, cross, star, and wavy lines. The idea is to shuffle the deck and then predict each card in turn. This type of format is typically done to test telepathic abilities. One person (the sender) looks at a random card, attempting to send the information to the other person (the receiver), who will then make their prediction. Because there is a one in five chance of simply guessing the correct card, someone would need to do significantly better than a 20 percent hit rate for the results to be considered meaningful.

In a typical research study using Zener cards, the sender and receiver would go through the deck of twenty-five cards numerous times, resulting in hundreds of trials; the more trials, the less likelihood that any significant findings are due to chance. Later, researchers shifted to using computer-generated symbols, which simplified many elements of the process, as well as providing much more control of the experimental design. Between 1935 and 1987, there were 309 studies examining both clairvoyance (seeing things in the present) and precognition (seeing things in the future) using computer-generated symbols or cards (Honorton and Ferrari, 1989). The combined results of these studies resulted in odds against chance of 10^{25} (10 million billion billion) to 1 (Radin, 2009). Another examination of forced-choice psi research was published nearly ten years later. In this meta-analysis, the researchers examined twenty-two studies published between 1935 and 1997. They reported clairvoyant effects with odds against chance of 400 to 1 and precognition effects with odds against chance of 1.1 million to 1 (Steinkamp et al., 1998). Clearly, variations of the Zener card method have been successfully used to measure psi abilities for decades. Perhaps this would be a good method for testing my own abilities.

Testing Myself

I didn't have a Zener deck of cards, but I did have a knockoff version that was gifted to me by a friend. He knew I was into all this "weird" stuff and gave me a board game that must have been sitting in his basement for twenty years. It was called Kreskin's ESP and was mostly geared toward using a pendulum to answer questions about the future. The game also contained a deck of twenty-five cards— five cards each of five different colored circles (red, orange, yellow, green, and blue). I decided to use these for my self-experimentation.

Since I didn't have anyone to run telepathy tests with on a consistent basis, I decided I would just shuffle the deck and attempt to read the cards with my psychic abilities. I would draw the top card, stare at the back of it, attempting to visualize the color of the circle on the other side. Once I made my prediction, I would flip the card to check my accuracy. If I was correct, I would place it to the right of the pile and if it was incorrect, I would place it to the left. At the end of the twenty-five cards, I would calculate my "hit rate" and then document it. I decided I would do this every day (or at least most days) to get a solid baseline of my psychic ability. As any day could be good (that is, accurate) or bad (few matches), it was important to have lots of trials over lots of days. This was the only way to get an accurate picture. However, I learned very quickly that there are certain advantages and disadvantages to using a deck of cards for this type of experiment.

Cards and Computers

When you have a known quantity of each card, and you see the results of each trial, it is very difficult for the logical, analytic part of your brain to get out of the way. It really wants to predict what will happen next. For example, if three of the first five cards happen to be yellow, some part of your mind is going to tell you, *There's no way the next card is yellow.* This natural inclination can get in the way of simply listening to what your intuition tells you. On the other hand, the fact that this predictive tendency is so strong provides a great opportunity to observe this process and learn to approach each card as if it were the first of the deck. Essentially, this practice becomes a psychic training exercise to learn the subtle differences between prediction and intuition.

To find a solution to this predictive problem, I started looking for a computer-based psychic test, similar to what had been used in some of the research studies described above. I didn't have to look far. One of the mediums in my case studies, Joanne Gerber, had a test on her website. It was designed very much like the Zener card approach although each trial was completely random. The digital cards could be numbers (one through five) or colors. Because the computer randomly selected one of the five options for each trial, there was no way to logically predict what "card" would be next.

Over the course of several months, I tested myself with the physical-colored cards, and with Joanne's website game, using both numbers and colors. I kept track of my performance and calculated my overall hit rate. After twenty-four runs using twenty-five cards each time (600 trials), my overall average was 24 percent. If the results were completely due to chance, I should have scored 20 percent. Not bad. When I analyzed a bit further, I found that I did the best using the physical color cards (26 percent) followed by the computer color cards (24 percent), then the computer numbers (20 percent). Because it is possible that the physical card success rate was inflated by my ability to predict, I suspect the computerized success rates are a more accurate metric.

Even though these results are likely to be significant, I felt they did not tell the entire story. When I plotted the results on a distribution chart, it was clear that there were far more runs at or above chance than there were below chance. This is important because it demonstrates a level of consistency that is not obvious when just looking at the overall hit rate.

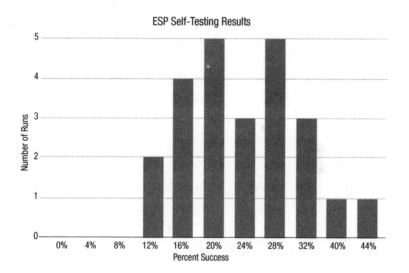

Of the twenty-four runs, thirteen had results better than chance (above 20 percent), five had results at chance (exactly 20 percent), and only six had results below chance (less than 20 percent). In addition, if you look at some of the extreme scores, you can see that there were times that I was able to get ten (40 percent), and eleven (44 percent) cards correct out of twenty-five. If these were high scores that occurred by chance, you would expect that there would also be comparable low scores—which there weren't. This was encouraging and provided more confidence than simply looking at the overall hit rate. Throughout the self-testing process, I also noticed some quirky patterns that would sometimes emerge between the target and the response, which may not be entirely random.

Streaks. Often, I would notice that I would get several correct in a row and then seemingly "lose it." On multiple sessions, I would have notes like, "first ten, getting 40 percent" or "got three of the first six and then went on a missing streak." This process felt like riding a wave. If I could tap into the natural flow and rhythm, it was effortless and automatic. Then something would shift, I would drop out of the flow, and have to reset my attention.

Correspondences. At times, I would notice strange prediction/response patterns that felt like they were meaningful. For example, there were sessions where every time I chose a certain color (let's say green), the actual card was blue. This observation seemed to happen most frequently with colors that were near each other on the color spectrum, so blue/green, green/yellow, orange/red. This seemed odd but made me wonder if somehow the colors were too close together for my more subtle perceptions to accurately identify the difference. Another pattern I noticed was the "one off" phenomena. Essentially, the color I selected was wrong, but ended up being the next card in the stack. For example, I predict yellow and the actual card is red, but then when I flipped the next card, it is yellow. Now, if this happened occasionally, it would be easily written off as a coincidence, but when it happens consistently, it starts to feel more meaningful. Is it possible that sometimes the mind is moving ahead in time?

Two Choices. Relatively often, I would be focusing on the next card, waiting for an intuitive hit, and I would see two colors in my mind, both seemed equally likely to be correct. More often than not, I would choose one and it ended up being the other.

I bring up these patterns only to point out that the reality of psi phenomena appears to be messy and complicated. It is not likely to be linear. If we truly are shifting our minds into a universal consciousness, it makes sense that the information received may be from a slightly different point in space or time. Perhaps this is why it is rare to find people who can consistently produce results on these tests greater than 30 percent.

I was curious if there was anything I could learn about the psychic mediums I had studied that might help my performance be a bit more consistent. What did their brain maps look like during a psychic reading? Was there some way I could help my brain do the same thing? Would this improve accuracy and/or consistency?

The Psychic Brain

One of the striking differences between Laura Lynne Jackson's brain during a mediumship and psychic reading was the type of activation in the occipital lobe (see Chapter 2). While the mediumship reading showed an increase of slow brain waves (and only slow brain waves), the brain patterns during a psychic reading showed something quite different. During the psychic reading, Laura's brain showed a significant increase of activity in the *fast* brain waves, particularly in the gamma frequency, which is the fastest brain wave we are typically able to measure.

After looking at a few more brains belonging to mediums, including Janet Mayer, Joanne Gerber, Angelina Diana, and Rebecca Anne LoCicero, it started becoming clear that in nearly every case, the gamma activity was increasing in the back of the head during psychic readings. Below is an analysis of Joanne Gerber's brain wave activity comparing her baseline EEG to the EEG during a psychic reading.

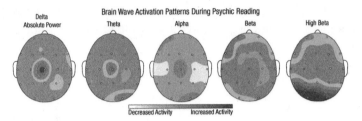

In this image, the darker shading represents significant increases, while lighter colors represent significant decreases of activity. Clearly, the strongest findings for Joanne were an increase of both theta (centrally) and high beta in the occipital lobes.*

As noted elsewhere, the increase of theta, may help with the ability to tap into universal consciousness. We can think of increased theta activity as the mind beginning to let go of its rigid grip on the external world and relaxing into a more creative, free, internal awareness. Experienced meditators in specific traditions often show patterns of increased theta and alpha activity, which is generally considered a deeper state of consciousness. In fact, the examples provided at the beginning of this chapter all occurred during sleep, which is most certainly a deeper state of consciousness. In addition to this increase of centralized theta, Joanne, and nearly all the other psychics studied, demonstrated a dramatic increase in gamma wave activity in the occipital lobes.

*Note: The system used in this analysis does not provide data for gamma, although it should be noted that high beta and gamma frequently overlap and are essentially next door to each other in the frequency spectrum.

Gamma Brain Waves

Gamma is the fastest activity that we are typically able to measure in the brain. Depending on the researcher or clinician, it can be defined as the brain waves occurring anywhere between 30 Hz and 110 Hz. This set of frequencies has gained a lot of attention in the last few years as researchers have connected it to a variety of different (and interesting) states of consciousness. On a basic level, increases in gamma are observed when there is increased activation. In fact, of all the EEG frequencies, gamma shows the strongest and most consistent relationship to glucose metabolism, a well-known indicator of brain activity (Oakes et al., 2004). Not surprisingly, increases in gamma are also associated with conscious attention and focus (which lead to increased activity). While these findings are interesting and useful in understanding how the brain works, these are not the reasons people get excited about gamma.

Gamma is also associated with states of flow or "being in the zone." Research has demonstrated that when experts in various fields are engaged in activities related to their expertise, you see significant bursts of gamma activity. This has been shown when musicians are listening to music, as well as when expert meditators are engaged in various forms of meditation (Lutz et al., 2004; Braboszcz et al., 2017). In these instances, the gamma waves seem to be involved with an effortless doing. Rather than thinking about the situation or analyzing it, the persons performing the activity are simply "tapped in" to the process. This is why people are interested in gamma. Who wouldn't want to live life in a state of flow, where you are able to process complex information at a high level seemingly without effort?

Gamma has also been implicated in various higher states of consciousness. For example, neurophysiologist Juan Acosta-Urquidi,

PhD, reported significant increases in this EEG band in subjects undergoing Kundalini activation; at the same time, the subjects reported being in a blissful "orgasmic" state which lasted several minutes (2019). Elevated gamma has also been observed during ayahuasca visionary states (Stuckey et al., 2005), and during lucid dreaming (Voss et al., 2009). These findings suggest that gamma's importance extends beyond simply identifying if the brain is active or not, it also seems important for expanded states of awareness. Given this understanding, it makes sense that we would see this kind of response in the psychic brain. And this activity is most obvious in the back of the head, the occipital lobe of the brain whose primary function is visual processing. To me, it makes sense that this area would be super-activated as the psychic is literally seeing beyond their normal visual capacity.

My next experiment involved exploring how I could increase gamma activation in my occipital lobes. I could use tech, but I also wondered if there was a meditative approach I could use that would work. Because I have a background in meditation, I considered which practices left me feeling activated rather than tranced out.

Third-Eye Focus

Throughout the several-month period that I was consistently working with the cards, I experimented with a variety of mental strategies to get better at the ESP games. For many of the early tests, I would close my eyes, and say something like, *Show me the next number,* and then go with the first number that popped into my head. Turns out this was not a terribly effective strategy. However, through some significant trial and error, I discovered a method that seemed to consistently result in a higher percentage of success. The first step in the process was to shift my focus to the center of my forehead. I could usually tell if this area was "activated" by a tingling sensation that would occur there. Then, I would imagine a window opening, allowing me to see into another layer of reality. I would intentionally let go of any real-world information or expectations and then just wait. I had to practice patience as all of my other strategies were based on the idea that the first thing that pops into my head is likely to be the best answer. This is apparently not true. Evidently, the first thing that pops into my mind is usually something tied to the mind attempting to predict based on what happened previously. All this automatic analysis is going on behind the scenes and, of course, masks the ability to tune in to what will happen next. By dropping these expectations and shifting the point of focus and then waiting patiently for a clear answer, my percentage increased significantly.

Third-Eye Associations

In Hindu and yogic traditions, the third eye is associated with the sixth chakra. In Sanskrit, this chakra is named *Ajna* which means "to perceive." The Ajna is a subtle energy center located in the center of the head, behind the forehead, either at eye level or slightly above. As noted by therapist and author Anodea Judith, PhD, the function of this chakra is to "see beyond the physical world, bringing us added insight" (2012). It is a subtle energy organ involved in psychic vision or clairvoyance (clear seeing). From this perspective, when the third-eye center is activated and functioning properly, it enables the ability to see beyond the limitations of the physical eyes.

In addition to the more esoteric descriptions, each chakra is connected to, or resonates with, a specific gland in the physical body. The ajna chakra corresponds to the pineal gland, a pine cone-shaped gland located approximately in the center of the head at eye level. This gland secretes melatonin, a hormone that is triggered by exposure of the physical eyes to certain light frequencies, and that influences our sleep/wake cycles. Interestingly, in some animal species the pineal gland/third eye connection is even more obvious, existing as an actual light sensitive organ on the surface of the head. It is also believed by some researchers that the pineal gland is responsible for secreting an endogenous form of DMT (dimethyltryptamine), an extremely potent psychedelic (see Strassman, 2000).

Realizing that there may be something to this third-eye business, I altered my morning meditation to focus on this area. I was again able to find that tingling sensation and found my energy increasing while just sitting there. In fact, I could also feel my heart rate increasing. I wondered if this is what the mystics and occult teachers are talking about when they use the term "raising your vibration." I know what my brain wave state is most of the time, and it is actually a little on the slow side—lots of alpha waves. I wondered if this practice (developing and opening the third eye) is actually resulting in increased gamma activity. Lucky for me, I can find out.

Third Eye and Gamma

I hooked myself up and ran a baseline EEG recording, then I recorded myself while I did two of the ESP games (twenty-five trials each). I video-recorded my screen so I could go back and review the exact points at which I was successful on the game. For the first run I scored 20 percent, exactly average. On the second run I scored 28 percent. After completing the tests, I watched the video, which allowed me to see both the ESP game and the EEG data simultaneously. I marked the time signature of the EEG at every point that I had a successful trial on the ESP games. I then went back and pulled out each of those successful moments, collapsing them together to create an EEG record of successful hits. I wanted to see if this data was any different from my unsuccessful trials, and if so, how.

When I ran a statistical comparison, it showed that there were significant differences in several bands, but especially in gamma and high gamma. In fact, when you look at the direction of the changes that took place, all the others showed less activity during the ESP hits. The only brain waves that increased during the ESP hits were high beta and gamma, and it was clearly focused in the occipital regions.

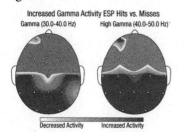

Increased Gamma Activity ESP Hits vs. Misses
Gamma (30.0-40.0 Hz)   High Gamma (40.0-50.0 Hz)

Decreased Activity   Increased Activity

When I looked at the size of the change, the analyses showed that gamma increased from 45 to 67 percent in the occipital regions. Interestingly, the change in beta activity showed a decrease in frontal regions, particularly on the left side. This part of the brain is very much involved with processing information and what we might normally consider "thinking." So, while my occipital gamma was increasing dramatically, my frontal beta was decreasing, suggesting that the ESP hits were not something that was happening as a result of conscious thought.

This was all starting to make sense. The third eye is involved in psychic vision, which is processed in the brain in the same areas that normal vision is processed, that is, the occipital lobes. When you engage in psychic seeing, you are likely to activate these centers, resulting in an increase of gamma activity. At the same time, it is necessary to shut down or quiet the logical, analytical, left-hemisphere dominant parts of the brain that essentially create noise in the system, making it difficult to see the more subtle information.

As a side note, while working on this chapter, I began practicing with the third-eye meditation again and got my highest score yet: 52 percent.

Try It Yourself

The best way to develop your psychic skills is to practice. You can purchase a deck of Zener cards online, but that is really not necessary. You can use a regular deck of playing cards and predict black or red. You could also predict the suite (clubs, spades, hearts, diamonds). You can also find apps for your smartphone or tablet with ESP games based on the same principle. I recommend *Are You Psychic* by IA Creative Software. This app allows you to choose from several different virtual decks of cards, as well as the option to have every trial be completely randomized or drawn from a deck of twenty-five cards.

Zener Cards

Practicing psychic games by using a deck of Zener-type cards can be very educational. By methodically working with these games, paying careful attention to your feeling state, taking notes, tracking performance, and being honest with yourself, it is possible to develop an understanding of how the psi process works for you—how to shift your attention and consciousness for optimal results. For myself, I was able to develop a subtle awareness that signaled to me when I was on the right track, which then helped guide me in later sessions. Here are some of my observations from several months of practice.

Format. My experience has been that every different format I use takes a bit of adjustment. The first few times I try a new deck of cards or a different computerized ESP game, I usually don't do very well. It is almost as if my brain must learn how to work with that particular system. As you start with a new approach or a new format, be patient and allow yourself some time to adapt.

The Feeling. Frequently, I have noticed a hard-to-define, felt sense, that would come over me when I know the correct answer. It is like a confidence or a clarity. When I have this feeling, I am nearly always correct in my prediction. When I do not have this feeling, the results seem random. During these nonconfidence trials, I could sometimes feel my mind trying to predict or extrapolate a pattern. If you can find this feeling and learn to recognize it, it may be possible to activate it intentionally.

Warming Up. At times, I have found that I do not do well at the beginning of a deck of cards. I might miss the first eight in a row before I settle into the session. For this reason, it may be useful to go through ten or so cards at the beginning as a warm up. If you consider psychic abilities as a skill, this idea makes a lot of sense. If you are an athlete or a musician, you most certainly warm up before a performance. It is an opportunity to find the feeling before it counts.

Decline Effect. This is a well-known observation in psi research and describes the tendency for people to do worse over extended testing sessions. Just as it may be important to have a bit of a warm up before beginning, it may also be important to keep each session contained to a certain time frame. It is not entirely clear why performance seems to drop over time, but it may be due to boredom and lack of engagement. Doing the same thing repeatedly may not present enough motivation to fully tap in. In addition to keeping your sessions contained, you may also want to mix it up. Try something different to keep it fresh and/or keep track of your results and all of the variables you notice as a way to increase engagement.

Third-Eye Meditation

Of all the mental approaches I tried, the one that seemed to work best involved "looking" through my third eye. I would do this by simultaneously directing my attention to the center of my brain and the center of my forehead. I would then relax my entire face and scalp, visualizing my brain and forehead as active and open. I usually knew this approach was working when I felt a slight tingling sensation on my forehead. Try the following meditation for a few weeks to see if this enhances your psi performance.

Third-Eye Meditation Script

Take a moment to settle the body. Relax any muscles that are holding any unnecessary tension.

Relax the breathing, allowing it to flow naturally and easily. No effort, just smooth, gentle, long, slow, and natural breathing. Feel the belly expand with each inhalation and the belly contract with each exhalation.

Now, draw your attention to the forehead, the space between the eyes or perhaps slightly higher. Imagine with each inhalation breathing purple light into this space. With each exhalation, let go of any stagnant or stale energy that may be in this area.

Breathing in electric violet light, activating your third eye.

Breathing out smoke and stale energy.

Keep the forehead relaxed.

Keep breathing, noting any sensations that might emerge.

Continue for several minutes.

Now, shift your attention and imagine breathing white energy in through the crown of your head. Image that with each breath, this white light energizes and activates your crown chakra.

Exhale any stale or toxic energy out of your crown, allowing this energy center to be open and receptive.

Continue for several minutes.

Now, imagine simultaneously breathing in from both the third eye and the crown. Feel these areas tingle with energy.

Imagine drawing energy in on the inhalation and visualize these energies meeting in the center of your brain in the pineal gland.

On the exhalation, sink this energy into the pineal gland, activating and energizing, waking up the third eye and your clairvoyant abilities.

Keep the attention in the center of the brain. What do you notice? What do you feel?

Continue this practice for several minutes.

Relax your attention and intention and let go.

A guided version of this meditation is available at www.psychic-mindscience.com.

CHAPTER FIVE

TELEPATHY AND AUTISTIC SAVANTS

WHEN I AGREED to be part of the docuseries *Signal,* directed by Ky Dickens, exploring the inner world of children on the autism spectrum, I was not expecting it to completely shake my world-view. My job was to examine the brain activity of autistic children that reportedly displayed telepathic abilities. It was hoped that this preliminary testing would help us decide if the topic was worth exploring further. I had always prided myself on being open-minded and accepting of the unexplained, but this project revealed my own limitations and areas for growth. Apparently, I had subconsciously believed that most psychic abilities were simply enhanced intuition, resulting in slightly better than chance predictive abilities. I wasn't sure what to do when I was faced with clear and direct evidence of flawless mind reading.

My friend and colleague, Dr. Diane Powell, who recommended me for the project, had been studying savant abilities in autistic individuals for over a decade. A savant skill is any kind of special talent that was not learned in a traditional manner. You might think of the scene in *Rain Man* where Dustin Hoffman's character Raymond Babbitt knows the exact number of toothpicks spilled on the floor without counting them.

These types of skills are rare and are generally only seen in certain people with neurodivergent disorders, such as autism. Currently, there are five commonly recognized categories of savant skills, that include art (e.g., hyper-detailed drawings); music (proficiency with musical instrument(s) often with little to no prior training); memory (recall of facts, events, numbers, etc.); calendar calculation (the ability to provide the day of the week for any given date); and mathematical calculation (fast mental arithmetic).

Dr. Powell's research has included reports of telepathic abilities in nonvocal autistic children from all over the world. In the same way that some kids on the autism spectrum can do amazing things like play any musical instrument expertly or understand advanced mathematics, Dr. Powell argues that the telepathic abilities she has witnessed, if scientifically validated, should also be recognized as savant skills. For this to happen, mainstream science would need to acknowledge that telepathy exists in the general population and that some autistic individuals have exceptional capability in this area. This does not seem likely to happen.

When I first became involved in the project directed by Ky, I had limited exposure to telepathy and even less exposure to kids with reported telepathic abilities. I wasn't at all sure what to expect or how this was going to go.

I was familiar with research looking at some people's ability to see or know things they shouldn't be able to see or know, but quite frankly, most of the results are less than awe inspiring. They are statistically significant, but not that impressive. For example, recall the Zener card experiments discussed in Chapter 4. As a reminder, this is a common ESP test that involves using a special deck of twenty-five cards with five different symbols on them. There are five cards of each design in the deck. The person being tested can either predict the next card to be turned over or a "sender" can look at the card while the other person attempts to read it from the sender's mind as the receiver, telepathically.

In a test like this there is a 20 percent chance of getting the correct answer by simply guessing (one in five). Most psychics doing these tests score between 25 and 30 percent, which is good. A few will score as high as 40 or 50 percent correct, which is amazing. What I witnessed with the kids in the docuseries pilot was 100 percent accuracy with a variety of tests, some of which were far more difficult than the exercise described above. This was positively unheard of, and I was determined to understand what was going on.

The first child I worked with on this project was Elisa,* a nonvocal thirteen-year-old girl on the autism spectrum. Having never met her, my initial curiosity was how she would respond to all of the people, cameras, and general commotion present at a film shoot. People on the spectrum are often highly sensitive to sensory stimuli and can become overwhelmed by new environments or changes in their routine.

Ky and her production team had spent time with Elisa and her family the night before and kept talking about what a joy she was to be around. They were not wrong. She smiled almost the entire time and would sometimes just break out in spontaneous laughter. Her

*Names and some other identifying information in this chapter have been changed to protect the privacy of the families involved in this research.

happiness was contagious and made you want to be with her. Elisa was accompanied by her mother, father, and cousin, who all exuded warmth and love. It was easy to see where Elisa got her temperament.

I discovered that Elisa had just started communicating with a letterboard a few months prior to the film shoot. A letter board, sometimes called a communication board, is simply a tablet with the letters of the alphabet and numbers from zero to nine on them. Some nonvocal people with autism can use this device to communicate by pointing to letters, spelling words, or using simple phrases.

Dr. Powell and Ky had conducted one or two Zoom meetings with the family prior to this in-person meeting and had a good sense of Elisa's abilities. I, on the other hand, had no clue what she could do.

The team decided we would begin with some simple tests with Elisa and her mother, Maria, the person with whom she has telepathic contact. Because there was already a lot of stimulation, we thought it was best to wait on the EEG part of the study. The idea was to let everyone get comfortable and introduce some of the telepathy tests in a more relaxed format.

The Random Number Generator Experiment

Before the first experiment began, Ky and her camera team moved through the room removing any reflective surfaces and covering the TV. They seemed to be taking extraordinary measures to minimize the possibility of any "cheating." I was observing from another room, watching on a video monitor. First, Dr. Powell fitted Elisa with a specially designed blindfold with extra foam around the sides to prevent any possibility of seeing anything (in this field this is called a *Mindfold*). Dr. Powell then opened her iPad to a random number generator (RNG) application. After selecting a range of numbers, this app randomly chooses one within that range. They decided to begin with a three-digit number. Dr. Powell selected the range from 100-999 and hit the start button. Nearly instantaneously the screen displayed the number 698. Dr. Powell angled the screen away from Elisa and showed the number to Maria. The turning of the iPad screen was an extra precaution. After Maria saw the number, Dr. Powell closed the iPad and removed Elisa's blindfold. At this point Maria placed one finger on Elisa's forehead. Immediately, Elisa began pointing to numbers on the letterboard: six, then nine, then eight.

My jaw hit the floor. I have witnessed many interesting and amazing things in my life, but I had never seen anything like this. The skeptical part of my brain didn't know what to do. Having observed the whole process, I could not see any way that cheating was involved. In fact, Elisa seemed to have absolutely no interest in the arrangement of the task *and* she had on a blindfold *and* the iPad was angled away from her *and* no words were spoken during the experiment.

They repeated the experiment, but this time Ky placed a solid black 2 x 5 piece of foam core between Elisa and Maria. The number this time was 956. Once again, after showing Maria the number, Dr. Powell closed the iPad, removed the foamboard and blindfold, and Maria placed her index finger on Elisa's forehead. Elisa pointed to the board: nine, five, six.

Dr. Powell proceeded. "Let's increase the random number to four digits." She changed the range on the random number generator from 1000 to 9999 and hit the button. The number that showed up was 4089. I calculated the odds. With a four-digit number, there was a one in 9,000 chance that Elisa could guess the number correctly and she did, except she didn't seem to be guessing. She knew the number, just like she knew the other numbers—without seeing them or hearing them. She appeared to be receiving the information straight from her mother's mind.

By now, my heart was pounding. Dr. Powell, on the other hand, was completely calm, as if nothing particularly interesting was happening. Later, she told me that she has been observing this kind of thing for years with kids from all over the United States. While I had already been converted to a "true believer," Dr. Powell maintained a skeptical attitude. Despite the impressive results, she indicated that we really needed a much more controlled testing environment before this could be considered proof of telepathy. Perhaps Maria was providing Elisa with some kind of subconscious subtle cues that signaled the correct answer to these tests. For example, what if the physical pressure from Maria's finger changed slightly when Elisa was getting close to pointing to the correct letter or number? While this seemed unlikely to me, Dr. Powell reiterated the importance of obtaining airtight scientific evidence for something as mind-blowing as telepathy. This reminded me of the famous saying from astronomer Carl Sagan, "extraordinary claims require extraordinary evidence."

For Dr. Powell, Elisa's demonstration, was part of a much larger scientific experiment. For me, it was a clear demonstration that we have capacities beyond what we might ever consider—if we can only learn how to use our extra-ordinary abilities.

Vocabulary Flash Cards

After several more trials with the number generator, the team switched to vocabulary flash cards. Each card showed an image and its name. Some of the cards were in Spanish, Elisa's primary language, and some were in English. Dr. Powell shuffled the two decks together and randomly selected a card from the deck.

The first image was a horse. She showed it to Maria using the same procedure as before. Elisa's blindfold was removed, Maria placed her finger on Elisa's forehead, and Elisa began pointing to letters on her board.

The first letter she pointed to was a C. On the sidelines, I was heartbroken. She was getting it wrong. I wondered what had happened.

She continued: A, B, A, L, L, O—*caballo*. Suddenly my college Spanish came back to me; *caballo* is horse in Spanish. Elisa repeated the exercise with an image of shoes, spelling the word *zapatos*. As the experiment continued, Ky had the crew pick a card and hand it to Dr. Powell to remove any possibility of cheating. It didn't matter. Elisa accurately identified word after word. How could this be possible? Only two explanations made any sense to me: either Elisa was looking through her mother's eyes or reading her mind.

Through the translator, Ky asked Elisa if she was able to see through her mother's eyes. Elisa wrote, "No, I can see everything." This seemed to suggest that she was receiving the information by some other means.

This hypothesis was confirmed when Maria later described how they discovered Elisa's telepathy. During the past two years, Maria had been teaching Elisa how to read through the Doman Method, which involves learning to read words from a flash card, rather than from phonics. Apparently, this approach is often easier for children with reading difficulties or special needs. In November 2021, Maria wrote some instructions for Elisa, and she followed them, which confirmed that she could read. After this development they got her a communication board and an iPad to help her express herself more easily. Elisa's parents began testing other areas of learning to see what Elisa was retaining. They began asking her some simple math problems which she answered correctly. They then began asking more complex math problems which she also answered correctly with no problem. Maria said that she initially thought maybe Elisa had some mathematical savant abilities, but when asked about her math skills, Elisa communicated that she couldn't do the math—she was reading her mother's mind. From there, Maria began doing various tests to see if this was true. For example, she would think of something, like a purple bag, and Elisa would write the Spanish words for purple bag on a piece of paper. I saw them do variations of this exercise during the filming. While cameras were being moved or readjusted and the researchers and director were busy planning the next set of experiments, Maria and Elisa would sit at the table patiently waiting. Maria would think of a color and then Elisa would point to one of six colored popsicle sticks on the table. Each time, Maria confirmed Elisa's choice. I could see that Maria was not looking at anything, leaving me with only one conclusion. The only way Elisa could have received the information was by reading her mother's mind.

At some point, the director, Ky Dickens, began creating new experiments on the spot. She picked up a book and randomly opened it, then handed the book to Dr. Powell and asked her to choose a word on the page. Dr. Powell chose a word and showed it to Maria; Elisa had on the blindfold. As soon as the blindfold was removed and Maria put her finger on Elisa's head, Elisa immediately and correctly spelled out the word on her board, *even if it was in English* (neither Elisa nor her mother spoke any English at all).

What Factors Affected Her Success?

Being the curious, questioning kind of people we are, Ky, Dr. Powell, and I started asking Elisa to try the telepathy games in new and different ways. Could she read her father's mind? What would happen if Maria didn't touch Elisa's forehead? What if they were in separate rooms?

We tested all of these variations (and more) and found some interesting limitations to Elisa's abilities. To maintain her 100 percent accuracy, the tests had to be with Maria, and Maria had to make physical contact with Elisa. When Elisa tried with her father, she was only partially accurate and only when they were in physical contact. While this may seem odd on the surface, Dr. Powell told us that reports of telepathic communication seem to be the strongest and most consistent between people that spend a lot of time together engaged in the same activities. For some individuals, physical contact seems to help them access the ability.

On an intuitive level, this makes sense. We are better at connecting with the mind of someone that we have an intimate relationship with; the closer the emotional connection, the better the psychic connection.

Synchronized Brains

This idea has been examined by simultaneously recording brain wave activity between twins, couples, and friends to determine if their minds are somehow connected and if the degree of emotional connection makes a difference.

The first known study of this kind was conducted by a pair of ophthalmologists in 1965 at the University of Philadelphia (Duane and Behrendt, 1965). The researchers were monitoring the EEG activity of identical twins who were in separate rooms. The researchers would ask one of the twins to close their eyes, causing a large increase in alpha brain wave activity in the back of the head. At the exact time this happened, the other twin, in a separate room, showed a mirrored effect; their alpha activity also increased at the same time.

In several studies to follow, researchers extended this approach by flashing a light in one of the participant's eyes, while their partner was in a separate room. This type of study is clever because the brain consistently reacts to a novel stimulus (like a light flash) with a burst of activity in the visual processing centers of the brain. Similar to the alpha brain wave study, when one of the pair was exposed to flashing lights, the brain waves of their partner would respond in a similar manner, as if they were also seeing a burst of light (Targ and Puthoff, 1974; Standish et al., 2003, 2004).

Getting a bit closer to addressing our specific question, researchers at Edinburgh University conducted a rather elaborate study to examine the impact of emotional connection on telepathic bonding. Their study included three groups: pairs of closely connected individuals (related), pairs of strangers, and randomly assigned individuals. Using a method similar to the light-flashing studies mentioned above, the individuals in each pair were monitored in separate rooms while one person from each pair was exposed to a light flash. Statistically significant correlations were found between the brain wave patterns of those who were closely connected. The strangers showed a positive but smaller effect, and the unpaired individual showed no correlations with lights that were flashed in an empty room (Kittenis et al., 2004). These results all support the idea that our minds are connected to each other in ways that can influence brain wave patterns and that these synchronizations are strongest when there is a close bond between the participants.

When I can get outside of my research brain for a second and trust what I witnessed with Elisa's family, the evidence for telepathy is clear. What I observed was an amazing, loving, gentle family that has done everything they can for their daughter. They weren't looking for these abilities, trying to profit from them, or trick anyone. They appeared to be as amazed as I was. Their primary wish for Elisa is for her to continue to develop her ability to communicate to support her quality of life.

Note: At the time of this testing, Elisa had only been communicating through writing and using the letter board for about five months. Elisa has trouble with fine motor control, so writing has always been difficult for her. At some point, Maria began instinctively putting a hand on Elisa while she was communicating, which seemed to help with writing, using the letter board, and telepathy. When Maria learned from Dr. Powell that other autistic children with telepathy skills could connect without physical contact, they began experimenting. Initially, Maria would put her whole hand on Elisa's cheek. Slowly, she began removing fingers, first with four, then three, two, and now she just uses one finger. They also discovered that it does not matter where Maria touches Elisa, whether it is her back, cheek, forehead, or shoulder. Elisa is now sometimes able to read her mother without touch, but it is inconsistent.

So, how does she do this? How can Elisa so clearly and accurately see what is in her mother's mind while the rest of us are mediocre (at best) at these kind of telepathy tests? This was my main question and I hoped to answer it by taking a peek at her brain wave activity while she was engaged in telepathy with her mom.

Bring on the EEG

After the first round of testing and filming, we took a lunch break to give everyone a chance to relax, visit, and reset for the afternoon. We decided to repeat some of the tests from earlier in the day, this time while collecting 19-channel EEG recordings of both Elisa and Maria at the same time. This was a bit tricky as neurodivergent people often have heightened sensitivities to sensory input, and doing an EEG requires wearing a tight-fitting electrocap on the head, having a specialized conducting gel squirted into the sensors on the cap, and connecting the cap on your head to ear clips.

Elisa handled it surprisingly well. After getting baseline recordings for both Elisa and Maria when they were sitting still doing nothing, we started the telepathy tests. We managed to get EEG samples while Elisa did the random number generator test and using the flash cards. The entire time the testing was happening I was furiously scribbling in my notebook, documenting what was happening and the time mark on the EEG. This allowed me to select the exact EEG segments that corresponded to telepathic activity and compare it to baseline, essentially answering the question, how did Elisa's brain change when she was reading Maria's mind?

My Findings

When I compared the eyes-open baseline to the telepathy experiments with the random number generator, I saw there was a significant increase of activity in Elisa's brain across all EEG bands. This increase was particularly strong in the right frontal lobes.

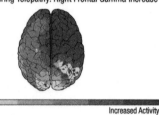

Elisa's Brain During Telepathy: Right Frontal Gamma Increase

Decreased Activity   Increased Activity

This is interesting as it is the opposite of what I found when examining Laura Lynn Jackson's brain during psychic tasks (see Chapter 2). When Laura did a psychic reading, her brain showed a significant *decrease* of activity in the frontal lobes, particularly on the right side.

Laura Lynne Jackson's Brain During Telepathy:
Right Frontal Gamma Decrease

Decreased Activity   Increased Activity

Clearly, this part of the brain is important for these skills, but it is unclear why Elisa's and Laura's brain activity would move in opposite directions. We might be able to gain a better understanding by considering the function of the right frontal lobe and the demands of the psychic skills examined.

First, it is important to note that Elisa and Laura were doing different tasks when their brain waves were recorded. Elisa was engaging in telepathy, which is an ability to transmit or receive information between minds, while Laura was using clairvoyance, the ability to gain information about a person or object through extrasensory perception (ESP). It seems likely that these abilities are related, yet not identical. Consequently, we cannot be sure that the difference in results was related to anything except that they were engaged in different tasks.

For a completely fair comparison we would need a scenario in which both people were doing the same type of psychic activity in the exact same way. Unfortunately, this is generally not possible as each psychic seems to have a particular set of skills and a particular way of doing things that works for them but may not necessarily work for others.

The Right Frontal Lobe

Even though Elisa and Laura were doing slightly different tasks, it is curious that both of their recordings showed significant changes in the same part of the brain—the right frontal lobe. Given the function of the frontal lobes, this seems like an important finding.

We typically think of the frontal lobes of the brain as responsible for the most complex types of thinking. They become engaged when we are involved in analysis, planning, or some form of decision making. The left frontal area is typically more involved in expressive language; this could be word-based thoughts, writing, or speaking. The right frontal area—the one we are interested in—is more non-verbal and involved in a more wholistic understanding as well as aspects of perspective taking and selecting among competing choices. In addition, the right frontal lobe has been specifically implicated in the ability to sustain attention.

So, why would Elisa's brain show increased activity in this region while Laura's showed a decrease? Again, it is possible that the type of task involved is important. For Elisa engaging in telepathy with her mother, it makes sense that she would need to recruit some of those right frontal lobe functions in order to tune in to what Maria was thinking. We can imagine that telepathy requires the activation of some of those right frontal lobe functions, such as perspective taking and directed attention. In essence, you need to be able to connect to what the other person is thinking while also screening out nonessential information (such as your own thoughts).

But couldn't the same be said for Laura and her psychic reading? Maybe not. Perhaps Laura is tuning in to something bigger than an individual and their thoughts. Perhaps she is tapping into a field of universal information. If this is what's happening, it makes sense that Laura might need to relax her attention, opening it up to allow access to more information. As I considered these findings, I became acutely aware that this was another instance in which there seemed to be a significant difference between the right and left hemispheres. Recall that nearly all of the mediums tested in Chapter 2 showed changes in the *right* frontal and parietal lobes. Perhaps this hemispheric asymmetry relates to a broader change in perspective that is often required for psi abilities.

Left Versus Right Hemispheres

While much of the functioning of the two hemispheres is similar and/or redundant, the left hemisphere has two distinct areas that are not generally present on the right side. The Broca's and Wernicke's areas, located in the front and rear of the left hemisphere of the brain, are critical to the expression and processing of language, respectively. This is important as language seems to be a central strategy for how humans define their sense of self. In her book, *My Stroke of Insight (2017)*, Dr. Jill Bolte Taylor, a Harvard-trained neuroscientist, talks about her experience of suffering a stroke in the left hemisphere. While she does not describe any mystical, transcendent, or psychic experiences, she does relate that the experience of losing language capability was extremely liberating. She was not constantly labeling, categorizing, and analyzing the environment, she explained; she was just experiencing. In a 2015 interview with Dr. Bolte Taylor on the NPR radio program, *TED Radio Hour*, she hesitated when asked if she would prefer to live in the left or right hemisphere. She explained that there was such a sense of peace and openness when the left hemisphere was silent, that she has missed that feeling since regaining the function of her left hemisphere.

The idea that the left hemisphere and its language functions complicate and potentially distort our experience of the world is supported by research on meditation. When the left hemisphere is engaged in meditation, it tends to take the form of a narrative and "mind wandering." This means the person is talking to themselves in their mind and/or creating a story about what they are seeing and experiencing. They are using language to explain and understand their situation. When the right side is activated during meditation, it is described as an experiential focus, an inhibition of cognitive elaboration in favor of a more direct sensory experience of the moment (Farb et al., 2007).

Thus, quieting the left hemisphere might allow us to open to additional information. For psi abilities, this shift toward the right hemisphere and away from the left hemisphere, allowing a more direct experience, may result in more accurate psi performance. There is evidence for this from a study conducted by Professor Allan Snyder at Australian National University (reported in Powell, 2009). Snyder used transcranial magnetic stimulation to disrupt the activity of the left frontal lobe of twelve neurotypical research participants. For a short time after the stimulation, subjects showed a significant improvement in their ability to guess the number of dots flashed on a computer screen. Essentially, the shift to right brain processing appeared to allow the participants to perceive nearly imperceptible information more accurately.

This and other data strongly suggest that the right hemisphere is connected to a broader spectrum of information, while the left hemisphere contains or limits the amount of information available to the conscious mind. We might think of the right hemisphere's processes as being more associated with the subconscious mind, and the left hemisphere's processes with the conscious mind.

Summarizing this data from the lens of extraordinary abilities, we might predict that quieting the left hemisphere (or shifting to the right hemisphere) is a central theme in the "tuning in" process of many psychics and mediums. In an interview with psychic medium Jeannine Kim, she offered a set of six tips to prepare for giving a psychic reading. The last of these steps was to, "Quiet the mind, and trust what you receive." I asked Angelina Diana about her process, and she described mediumship as "getting out of the way" and "opening to my intuition." This is a similar description to what occurs when we shift brain activation from the left hemisphere to the right. Janet Mayer told me that she would notice her "monkey mind" (left brain) often "running wild" at the beginning of her meditation practice. She described keeping a pad of paper and pen nearby so she could quickly jot down the logical brain ideas and thoughts that were streaming through. This often allowed her to then let go of those details and more fully tune in to her Higher Self (right brain). With practice, this became less necessary, she said.

I shared these left versus right hemisphere differences with Laura Lynne Jackson to see how they fit with her experience. She told me that she prefers to do psychic readings over the phone with her eyes closed. In this environment, she says, she is able to tap into a sense of oneness, a connectedness between all of us, and a force that is much bigger than the small self that we call Laura.

When she talked about this process it sounded as if she is able to release concepts of self and shift away from limitations created by the logical brain. She described the ability to see the person she is doing a reading for and all of their connections to other people. In this mental or etheric space, Laura can follow these energy cords to learn more about their relationships. In other words, she takes a more right-brained approach—less logical, less linear, and less verbal than the type of processing that tends to happen in the left hemisphere of the brain, where we are trying to figure something out.

These interpretations of how Elisa's and Laura's brains work might suggest that psychic skills require not one specific mental shift but a shifting of one's *normal* attention. In some cases, it will be important to narrow attention and screen out all other information. In other cases, it may be more important to relax attention to tune in to a broader field of information. These differences may depend on the specific psychic task as well as how your brain functions outside of psychic practice.

Clearly, some people have a natural propensity to shift attention in such a way as to encourage psi-related information. Maybe they are wired that way and have a form of mental flexibility that allows them to more efficiently navigate between left and right brain functioning. What about the rest of us? Are we doomed to a life of psi-lessness? How can we learn to either focus or expand our attention, connecting to information outside of our own mind? I believe the answer is practice, practice, practice.

Practice, Practice, Practice (with and without Help)

Just because certain psychic skills don't feel natural or your brain doesn't automatically shift into higher states of consciousness, doesn't mean they are impossible. It means you have to retrain your brain.

Believe it or not, we already do this all the time. In fact, the brain is constantly shifting and changing its connections and functioning based on what we ask it to do. This is the basic idea behind *neuroplasticity*. In the field of neurobiology, there is a famous saying attributed to the late psychologist Donald Hebb: "Cells that fire together, wire together" (Hebb, 1949). This tells us that brain areas that are activated at the same time during a specific task become linked. They develop stronger, faster, and more efficient connections, making this shared activation easier in the future. This is how states become traits; it is how learning takes place. If you repeat something enough times, it becomes hard-wired into the system, changing not only the structure but the function of the brain.

This process has been repeatedly demonstrated through brain imaging studies of meditation. Hundreds of articles from dozens of research labs have asked the question, "What happens in the brain with consistent meditation practice?" Turns out, it depends on what kind of meditation you are doing. Different types of meditation impact the brain differently and consequently affect different cognitive, psychological, and emotional skills. The brain changes in predictable ways based on what you ask it to do.

In one large and complex study, researchers assigned participants to one of three meditation groups, each teaching a different style. One group learned a focus style of meditation, another learned mindfulness, and a third learned open-heart practices. (Note: In an effort to use consistent language throughout this book, the meditation names used in the research study have been translated to the matching neuromeditation styles.) Each group practiced their assigned meditation style for three months and completed a variety of tests before and after, including an MRI to determine if any parts of the brain changed in response to their meditation practice. After each three-month cycle, the meditators would switch meditation styles. Over the course of nine months, all participants trained in all three styles. When the researchers examined the data, what they found was that each style of meditation grew specific parts of the brain and affected specific skills. The focus meditation resulted in growth in the prefrontal cortex and enhanced attention skills. The mindfulness practice influenced frontal and temporal regions of the brain and led to increased perspective-taking. The open-heart practice led to changes in the *insula*, a part of the brain involved in the experience of emotion and empathy, as well as increased scores on a scale measuring compassion (Valk et al., 2017).

This research and other similar studies demonstrate that the way you direct your attention and intention matters. It also shows us that we can adapt beyond our current programming, training the brain toward our desired goals.

Meditation for Psychic Development

In a recent interview with psychic medium Joanne Gerber, I asked her how she taught her students to get into the right mental state for psychic and mediumship readings. She replied, "mostly meditation." This should not be surprising as the connection between psi abilities and meditation has a long history in both Buddhist and Yogic/Hindu traditions. In these traditions, dating back thousands of years, it was well known that you could gain special abilities through the practice of meditation and yoga. These abilities, called *siddhis*, include psychokinesis (the ability to move things with the mind), clairaudience, telepathic knowledge, retro-cognitive knowledge (knowledge of a past event that could not have been learned), and clairvoyance (Seidlmeier et al., 2012).

The influence of meditation on psi abilities has also been established through scientific methods. In his book, *Real Magic*, Dean Radin, PhD, chief scientist at the Institute of Noetic Sciences, reported six conclusions that can be drawn from thousands of psi research studies. One of these conclusions states, "Psi effects are stronger during non-ordinary states of consciousness, such as during meditation" (Radin, 2018, p. 184).

Here is a brief example of some of this research evidence:

Researchers at the Institute of Noetic Sciences surveyed more than 1,000 meditators about their experiences with psi. The results indicated that approximately 75 percent indicated an increase of meaningful synchronicities in their life which they attributed to their meditation practice. Almost half reported sensing "nonphysical entities" during meditation, and a third reported some kind of experience with clairvoyance or telepathy (Radin, 2018).

Penberthy and colleagues (2020) compared a group of subjects who had signed up for a two-week meditation retreat or course to those with little or no meditation experience. Comparisons between the groups revealed that the subjects reporting a history of meditation practice and higher scores on a mindfulness scale were the same subjects most likely to report a history of paranormal and psi beliefs and experiences. The meditation group also performed significantly better on a task asking them to "guess" how many objects were in a jar that was displayed for only a brief time.

In a study published in the *Journal of Scientific Exploration*, researchers explored whether participants could significantly influence the output of a random *event* generator (REG) (versus the random *number* generator mentioned earlier). REGs are devices that do exactly as their name implies; they generate random events, usually in the form of 0's or 1's. With enough data points, you will expect there to be a zero 50 percent of the time and a one 50 percent of the time. In psi experiments with REGs, the idea is to see if people can influence these ratios through direct mental influence. Collesso and colleagues (2021) brought subjects into their office and assigned them to one of two groups. The first group was a control condition. After completing a variety of surveys, the control group was asked to influence the output of an REG by either increasing or decreasing the probability of a specified output (for example, creating more ones or zeroes). The second group did the same procedures, but also participated in a brief relaxation, meditation, and visualization exercise believed to be useful in changing the REG output. Consistent with their hypotheses, the group that engaged in meditation prior to the experiment was statistically more successful in shifting the REG in the desired direction.

Meditation as Mediator

If we compare the state(s) of consciousness involved in psi versus our normal waking state(s), we see how meditation practice might serve as a mediator between the two. With very few exceptions, most of the time our minds are running around in circles, obsessed with details, worries, memories, problem solving, and decision making. In fact, we are taught and trained through our society and culture that this is the appropriate and mature way to exist in the world. Unfortunately, this normal state is the antithesis of a psi-conducive state of consciousness, which requires openness, creativity, relaxed attention, trust, and the ability to listen.

Meditation is a tool to help us make this transition; to learn to relate differently to our everyday thoughts; to intentionally direct our attention and increase awareness of more subtle cues from our internal environment. In my earlier book, *Meditation Interventions to Rewire the Brain*, I defined meditation as "a systematic mental training designed to challenge habits of attending, thinking, feeling, and perceiving" (Tarrant, 2017). From this perspective, meditation is viewed as mental training. It is skill building. It is practice. When you consistently meditate, you are increasing your mental flexibility and developing the ability to navigate your internal landscape. As we saw above, the style of meditation you practice directly influences which parts of the brain become engaged (or disengaged) and which mental skills are influenced. It makes sense then, that different styles may also influence specific psychic skills.

Considering what we observed in the brains of both Elisa and Laura Lynne (as well as many other psychics and mediums), it seems that it might be helpful to practice meditation skills that exercise the right frontal lobe and/or quiet down the left hemisphere.

Try It Yourself

How you direct your attention and your intention during psychic states appears to be important to success. In the case studies above, we saw examples of both decreased and increased activation of the right frontal lobe. This suggests that it is important to develop flexibility and explore various forms of attention to discover what works best for you. The practices below can help with this skill development.

Focus Meditation

This style of meditation involves the right frontal lobe and includes any approach that requires you to direct your attention to a single target (breath, mantra, image of Buddha) while minimizing any other thoughts. When the mind wanders (which it will), the task is then to become aware of the mind wandering and gently, patiently, and lovingly, escort your attention back to the target. That's it. While it sounds simple, if you have ever tried this type of meditation, you know that it is extremely difficult. Focusing on one thing is boring and the mind will quickly become distracted, looking for anything to entertain itself. Remember, the point of a focus meditation is to direct attention toward a single target without becoming overly caught in other thoughts. The target could be internal (breath, third eye) or external (candle flame, image of goddess). The meditation below includes a few strategies to help you develop and explore your attentional capacity.

Three-Strategy Focus Meditation

This ten-minute meditation will explore three connected strategies to help you keep your attention on the breath. If you find yourself struggling with any part of the instructions, you can always shift back to an earlier part of the practice. Find your sweet spot that does not require too much effort.

Begin by tuning in to your posture. What would it look like if you were sitting in a state of alert relaxation? Keep the back straight, sitting up straight and tall but without any unneeded tension in the body. Keep the head perfectly balanced on the neck and torso, making sure that you are not looking up or looking down. The eyes can remain slightly open or closed if that is more comfortable.

Shift your attention to the breath. Allow your breathing to be relaxed and natural. Choose a way to focus on the breathing by noticing bodily sensations. Maybe you focus on the movement of the belly, expanding with each inhale and contracting with each exhale. Maybe you focus on the feeling of the air moving through your nostrils. Cool air moving in with each in breath. Warm air moving out with each out breath.

In general, it is best to keep the mouth closed, breathing through the nose, but if this is difficult, feel free to adapt to whatever form of breathing works for you. Invite the breathing to be slow and deep. Now allow a slight rushing sound to accompany each inhalation and each exhalation. This may remind you a bit of snoring or the sound of the ocean. Direct your attention to the sound of the breathing in combination with the sensations of breathing in the body.

Now, if it is comfortable, continue the deep breathing with sound, noticing the sensation in the body, and add a simple mantra to accompany each breath cycle. We suggest saying to yourself, *I am focused* during the inhalation and *alert* during the exhalation, but feel free to choose a different phrase that will help you stay connected to the intention of this practice.

Continue this practice on your own for the next few minutes, using the sensations, the sound, and the mantra to assist you in staying engaged with the practice.

You can find a recorded version of this meditation by visiting www.PsychicMindScience.Com.

Focus Meditation Tips and Tools

Having taught meditation for the better part of my life, I have figured out some tips and tricks that can help in the development of a focus meditation practice.

Start Slowly. A common mistake when beginning a new meditation practice is assuming that you have to meditate for thirty or forty minutes to achieve any benefit. This attitude is often a setup for failure. If you are unable to meet this arbitrary time limit, you might assume you can't meditate or that it isn't for you. The reality is that meditation is a skill and takes time to develop. Begin slowly, maybe only practicing for three minutes and gradually increasing the amount of time a few minutes each week. Set yourself up for success and let go of ideas about what meditation should look like.

Include Movement. If you find that seated meditation is difficult due to excessive mind wandering or trouble staying awake, you can add movement to your practice. This should not be something that is overly complicated as that will also take you out of the single-focus meditative state. In general, it is best to use a simple, repetitive movement that you can coordinate with your breathing. There are many qigong exercises that fit this role. For example, you can simply raise both arms above your head during an inhalation and bring them down the centerline of the body on the exhalation. You are still focusing on your breath, but now you are using your body to help you stay engaged. This can be done from a standing or seated position. There are a few examples of focus-style movement practices on our website, PsychicMindScience.com.

Body Posture. Most meditation teachers and books will recommend sitting in a stable posture, often with the legs tucked under you or in a lotus-type position, with the spine erect, shoulders relaxed, hands in a particular *mudra* (fingers positioned in a certain way), and chin slightly tucked. These suggestions make a lot of sense for many reasons, but they do not account for individual differences. At the NeuroMeditation Institute, we emphasize experimenting with a variety of postures to find the one that is most likely to facilitate the desired state of consciousness and be comfortable for you. In this case, we are hoping to be alert, focused, and relaxed. Well, what does that look like in your body? Because we are a mind/body, if you influence one, you influence the other. What happens if you sit in a chair? Or lay on the ground? Or stand up? Meditation is a state of mind, not a posture. Listen to yourself and discover what works best for you and then honor that wisdom.

Layering. We have already established that a focus meditation can be pretty boring and simultaneously mentally challenging. However, if you combine a few techniques together, it can add enough interest and stimulation to make the practice much more engaging. For example, rather than just watching the belly expand and contract, you could also count the breaths. Inhale, exhale, that's one. Inhale, exhale, that's two. In general, it is best to only count to ten and then return to one. This prevents competition by somehow trying to get to one hundred or one thousand. If you space out and lose track, just return to one. This strategy is adding a cognitive

element to breath awareness. You could also add movement as suggested earlier. Inhale, the arms go up, Exhale, the arms go down,
that's one. Now you have breath awareness, a cognitive element,
and movement. You get the idea. You can also add visualizations, or
energy sensing. In general, there is a limit to how many things you
can add before it becomes stressful or distracting, which is the
opposite of what you're trying to achieve. Experiment for yourself
and see what works.

Be Consistent. Repetition is how the brain changes. If you want
to build a new skill, if you want to strengthen your brain, you must
use it. To get the most from any style of meditation, it is important
to develop a consistent practice. Again, think of it like any skill—to
get better, you have to practice.

Quieting the Left Brain

In addition to practicing skills to activate or quiet the right
frontal lobe, it is also important to practice downregulating the hyperactive, story-telling tendency of the left brain. This can be accomplished by practicing both quiet mind and mindfulness forms
of neuromeditation.

Mindfulness Meditation

This style of meditation relaxes attention and quiets down the
frontal lobes, leading to a more open, spacious sort of attention.
While many programs and authors are currently using the term
mindfulness as an umbrella to include all types of meditation,
I am using it in a much more specific way. I like the definition
proposed by Jon Kabat-Zinn, founder of Mindfulness-Based Stress
Reduction. He defines mindfulness as "awareness that arises
through paying attention, on purpose, in the present moment, non-
judgmentally" (*Mindful*, Jan. 11, 2017). It is a calm awareness of

your thoughts, bodily sensations, feelings, perceptions, and even of consciousness itself. It is the process of observing what is happening right now, in this moment, with curiosity, yet without attachment. It is witnessing and accepting things as they are, without grasping for things we like or pushing away things we don't like.

When described this way, it becomes clear that mindfulness is a specific type of attention that is different from, say, focus. In a focus style of meditation, our attention is directed toward one thing, to the exclusion of all else. In a mindfulness style of meditation, the attention is open, allowing anything and everything to enter awareness, making it well-matched to support and facilitate psychic development. How can we practice this?

First, I think it is important to note that mindfulness can be both a formal and informal practice, and that both are important. A formal practice is what generally comes to mind when we hear the word meditation. This might be sitting on your meditation cushion every morning and watching your thoughts float by like clouds across the sky, or it might be attending a mindful yoga class, or participating in a meditation group at your local wellness center. All of these examples describe times (and spaces) dedicated to cultivating a mindful attitude. This is how we build our mindful muscles. Think of formal practice as your mindfulness exercise routine—going to the mental gym to build your mental muscles.

We may also extend this type of awareness to our everyday lives. This is what is meant by informal practice. For example: can you be aware of your thoughts, feelings, and bodily sensations while you are sitting at a stoplight? While grocery shopping? Walking the dog? Having a conversation? This third-person awareness of the present moment creates the capacity to notice what is actually happening, rather than focusing on worries and concerns related to what has happened in the past or what might happen in the future. The idea with an informal practice is to create a flexible attention. To increase the capacity to shift into a mindful type of awareness when you might ordinarily be in autopilot mode. Here are a few tips and practices you might try.

Slow Down: By definition, mindfulness means paying attention in the present moment. This is impossible if you are going a thousand miles per hour. Experiment with ways in which you can slow down: drive slower, turn off your phone, take leisurely walks, take a break during your workday, take a nap, relax your breathing, spend time alone, meditate.

Daily Mindfulness. Choose a daily activity—something that you normally do on autopilot without much conscious awareness, such as brushing your teeth, walking the dog, doing the dishes, or eating a meal. For the next week, intentionally pay full attention to what is happening in that moment. If you are cleaning your house, see if you can just clean your house without thinking of everything else that needs to be done, without mentally complaining, without fantasizing about your next vacation—just clean the house. Notice sensory information. What do you see, hear, feel, smell? Notice any bodily sensations. What do your muscles feel like as you move through the house and engage in various activities? Notice how shifting your attention changes the experience.

Formal Mindfulness Meditation. Spend a few moments allowing the body to settle. Feel the weight of the body being supported by your chair or meditation cushion. Let go of any unnecessary tension in the body. After you begin to feel settled and relaxed, observe with all your senses. What do you hear? What do you see in your mind's eye? What emotions are present? What bodily sensations do you notice? What thoughts? Whatever sensations, events, or thoughts seem to have the most salience, simply allow them to be present. Without judgment, gently observe and notice, recognizing whatever has shown up as a movement of the mind. Allow it to naturally pass. There is no need to push away experiences that are negative. There is no need to cling to experiences that are positive. Invite an attitude of openness and curiosity to each experience. Each experience is impermanent and will naturally move through your awareness like clouds across the sky. Simply observe, allow, and let go.

Nature. Find a spot in nature where you can sit comfortably for fifteen to thirty minutes. You can also find a spot that is indoors with a view of nature or use an immersive nature scene in virtual reality (www.tryHealium.com). Whatever the location, make sure that you are comfortable. Spend a few moments settling in, becoming centered and grounded, then allow the breathing to naturally slow, and invite tight muscles to relax. When you feel settled, tune in to the environment. Without actively scanning, allow the sensory experience of the environment to wash over you. What do you see? Hear? Feel? Smell? Notice how the sights, sounds, and tactile sensations impact your thoughts, feelings, and body. As much as possible, do not seek with the senses; simply "be" with whatever is present.

Movement. Practices such as qigong, tai chi, and yoga can bring an element of movement into the meditation. However, it is also important to recognize that mindfulness is about your attention, not the activity. You can just as easily do yoga mindlessly as you can mindfully; it all depends on your attitude. If you are fully present in the moment without additional commentary or judgment, then it is mindfulness. Movement practices can facilitate this type of awareness by providing specific and direct body-based sensations. Noticing alignment, connecting with the heavy, grounded feeling in the legs and feet, relaxing the muscles, minimizing effort, and engaging the desired muscle groups are all body-based sensations that can be directly observed in these practices. In addition, you can also notice what the movements and postures elicit. Do any feelings arise? Thoughts? Are you concerned with impressing the teacher? Are you concerned with impressing the other students? Are you competitive? Are you judgmental of yourself or others? All of these can be observed in the present moment as an act of mindfulness.

Botanical Blends

If your nervous system is chronically stressed or overactive, it may be difficult to quiet the mind. In these cases, it can be helpful to work with certain types of nonpsychedelic mushrooms and herbs to assist in relaxing the nervous system and steadying the mind. At the NeuroMeditation Institute, we have consistently found that people respond positively to reishi mushrooms, valerian root, chamomile, California poppy, and passionflower. All of these herbs and mushrooms have been shown to increase relaxation and create a mild sedation that generally results in a slowing of internal thoughts. The Quiet Mind blend was created by Evolved Mushrooms for the NeuroMeditation Institute to support our students in their ability to enter a meditative state in which there are minimal thoughts.

When I first began working with this blend, our research team tested it with a few volunteers to make sure it did what it was supposed to do. We measured brain wave activity at baseline and then while listening to a short, guided, quiet-mind meditation. After a short break we invited them to consume 1.5 ml of the Quiet Mind tincture and then do the meditation again. Below is a comparison of the changes in gamma brain waves during meditation, and the meditation plus tincture. Darker colors represent increased activity while lighter colors represent a decrease of activity. Notice that the increased gamma that accompanies the tincture is almost entirely in the right hemisphere, whereas the decreased gamma is almost entirely in the left hemisphere.

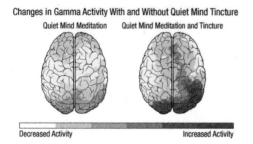

Changes in Gamma Activity With and Without Quiet Mind Tincture

Quiet Mind Meditation   Quiet Mind Meditation and Tincture

Decreased Activity   Increased Activity

One of the volunteers testing the blend noted, "With the tincture my mind became dark, completely empty, but not in a scary way. It was amazing." This is consistent with the research from earlier in this chapter showing that a shifting of activation to the right hemisphere is associated with a more direct sensory experience of the moment without the addition of analysis, judgment, or storytelling. In subsequent testing we have found that this blend appears to be particularly helpful for those with a busy brain. We have also observed that the dosage makes a significant difference and appears to be individually determined. You can find out more information about the Quiet Mind and other blends at www.neuromeditationinstitute.com.

CHAPTER SIX

ZAPPING THE BRAIN

GIVEN OUR EXPLORATION thus far, it appears that there are a handful of brain regions that are consistently associated with spirit communication and psychic abilities. These include the right parietal lobe, the right frontal lobe, and the occipital lobe. I began to wonder what would happen if we artificially stimulated those areas. Could I use technology-based interventions to help "wake them up"? Was it possible to utilize neuromodulation techniques to speed up the psychic development process? Rather than reinventing the wheel, I considered previous attempts to explore something similar.

The Right Parietal Lobe

One of the most consistent brain regions implicated in both psychic skills and spirit communication is the right parietal lobe (RPL). This region appears to play an important role in the ability to tap or tune into information beyond the self. When this part of the brain is "behaving" and doing its normal job, we have clear boundaries; we experience ourselves as separate individuals, contained in our bodies. We know where and who we are, and we see that as separate and independent from others and the larger world. When we disrupt the functioning of this region through injury, meditation, or altered states, these self/other boundaries become weakened and open the possibility of connecting to energies beyond the self. In that space, we are not isolated and independent; we are interconnected, allowing us to move beyond our small sense of self and connect with those around us, or even to something greater.

Not surprisingly, many of the psychics and mediums examined show a clear pattern of shutting down or disrupting activity in this part of the brain when they are doing their psi thing. Sometimes this appears similar to seizure activity in the brain. Other times it looks like a substantial increase of slow brain wave activity; either way, this part of the brain is no longer engaged in its usual activity.

Zapping the RPL: A Brief History

In the first study to explore brain stimulation and anomalous experiences in 2002, Blanke and colleagues used implanted electrodes in the brain of an epileptic patient to examine the impact of electrically stimulating different parts of the brain. When they stimulated the *angular gyrus*, a specific area in the right parietal lobe, the patient reported experiences that seemed to mimic an out-of-body experience (OBE).

When the researchers used lower intensity stimulation, the patient described feeling like she was "sinking into the bed" or "falling from a height." When they turned up the intensity, the patient described a full-blown OBE. In her words, "I see myself lying in bed, from above, but I only see my legs and lower trunk." The researchers repeated this process several more times, each time with similar results.

In a related study, researchers used *repetitive transcranial magnetic stimulation* (rTMS) to disrupt the functioning of the RPL (Uddin et al., 2006). rTMS uses an electromagnetic coil placed just above the scalp to send magnetic pulses into specific regions of the brain. This technique has recently been used as a treatment for depression, PTSD, obsessive-compulsive disorder (OCD), Tourette syndrome, and various other movement disorders. The repetitive part of the name refers to the stimulation sent as a recurring pulse, turning on and off at different rates. In general, faster rates (10 to 20 Hz) tend to be more stimulating, activating the brain region, while slower rates (1 Hz) tend to be more quieting, shutting down the brain regions beneath it. This approach is easier to implement than the previously mentioned study, as you do not have to directly implant electrodes in the brain. In that way, the rTMS approach is considered noninvasive.

In this second study, the researchers pulsed 1 Hz into the RPL and then presented the research subject with a series of images. Some were morphed images of themselves, and some were morphed images of friends and colleagues. The participant's job was to identify if the image they saw was them or someone else. After the rTMS treatment, participants performed significantly worse on this task. When the researchers tried the same approach on the left parietal lobe (LPL), this ability was not affected.

These two case studies suggest that stimulating or disrupting the RPL with either an electric or magnetic pulse can lead to out-of-body experiences as well as an interruption of the ability to distinguish self from other. These results are promising and seem to be pointing us in the right direction, yet they don't necessarily suggest an increase in psychic or spirit-communication skills.

The Temporoparietal Junction (TPJ)

If we extend our examination of the right parietal lobe to include the adjacent brain real estate, we find ourselves in an area called the *temporoparietal junction* (TPJ). This area serves as the intersection of the temporal lobes and the parietal lobes and is also frequently implicated in anomalous experiences. In a paper titled *The Out-of-Body Experience: Disturbed Self-Processing at the Temporo-Parietal Junction* (Blanke, et al., 2005), the authors reviewed ten different published case studies of neurological patients experiencing an OBE, covering a total of thirty-four patients. The vast majority of these cases involved the right hemisphere focused on either the temporal lobes or the TPJ. While the temporal lobes haven't been prominent in the case studies I have presented, they have been referenced several times (see Joanne Gerber, Chapter 2). The work of Michael Persinger indicates that this area of the brain may also be an important target for some of our brain stimulation experiments.

The God Helmet

Michael Persinger and his colleague Stanley Koren spent decades exploring whether certain kinds of brain stimulation could be used to induce an OBE, and experience noncorporeal entities or mystical states. This wasn't their original intention, but rather a happy accident. Initially, they thought they might be able to stimulate creativity or inspiration by disrupting the usual interconnectedness of the left and right hemispheres. Their idea was to apply a weak magnetic field to the temporal lobes. As soon as they started this program of research, participants began reporting strange experiences, including the feeling of a presence, visions, or some other meaningful spiritual experience. With some of Persinger's protocols, up to 80 percent of the participants described something that might be considered "paranormal" (St.-Pierre and Persinger, 2006). For this reason, their device soon became known as the "God helmet." While it was not the original intention, targeting the temporal lobes may be a useful approach. Other research has shown that people with temporal lobe epilepsy often report heightened spiritual experiences, including OBEs. These unexpected experiences shifted Persinger's research focus. Through experimentation he found that there are several different stimulation approaches that seem to be effective with somewhat different outcomes. Apparently, stimulating the right temporal lobe in a particular manner often led to the feeling or sense of a presence oriented toward their left side. Doing a similar stimulation pattern on the left temporal lobe or on both temporal lobes also consistently led to the feeling of a presence, but its location varied, depending on the person; sometimes it was to the right, sometimes to the left, and other times it was above or behind the person wearing the helmet (St.-Pierre and Persinger, 2006).

Persinger's approach had another advantage. Rather than using high-powered magnetic stimulation like a traditional rTMS device, Persinger was using a lower-power magnetic stimulation device. While exposure to high-power rTMS has to be limited for risk of overheating the brain, low-power rTMS carries a low risk and can be applied for longer periods of time. This makes it much safer and easier to use for psychic development (it's also a heck of a lot cheaper and accessible).

As my work as a psychologist involves neuromodulation, I was familiar with low-power rTMS. In fact, I own several systems that have this capability and the flexibility to program them in targeted ways. Historically, I had used this tech to disrupt rigid patterns in the brain prior to neurofeedback. The idea was that it could be helpful to "nudge" the brain out of its stuck spots, following up with neurofeedback to tell the brain what you want it to do, so to speak. I was excited to try this approach for psychic development but wasn't quite sure how to start. Even though Persinger and others have published some information on their approach, they also kept some of the details of their protocol protected as a trade secret.

As I began thinking about the best approach for "zapping the brain," I consulted again with Dr. Diane Powell, a friend and psychiatrist, who has been studying psychic phenomena for decades. In her book, *The ESP Enigma*, she explored some of the research discussed above and suggested that unusual activity in the angular gyrus and nearby areas appears to facilitate access to psychic information (2009). She also suggested that the brain stimulation approach most likely to lead to interesting results would involve both the right angular gyrus *and* the temporal lobe(s). I wondered about combining the two approaches but wasn't sure whether to stimulate one temporal lobe or both.

Even though the data from the angular gyrus clearly points to the right side, much of the research with the temporal lobes implicates the left side. For example, as reported in *The ESP Enigma*, Willoughby Britton found that 22 percent of people with a history of near-death experiences (NDEs) demonstrated seizure-like activity in the left temporal lobe (2009). However, a study by Tedrus and colleagues (2023) found that people with right-hemisphere epilepsy (that is controlled by medication) report a higher frequency of spiritual experiences compared to those with left-hemisphere epilepsy. Perhaps the right and left temporal lobes are involved in different aspects of altered states, or perhaps it depends on the person.

Self as Guinea Pig

Each time I try a new technique, I begin with myself. I decided to use a consumer-grade rTMS device called the NeoRhythm, which looks like a heavy plastic headband. This device has four coils in it and is controlled through a Bluetooth connection to a smart phone or iPad. The design makes it easy to shift the positioning on the head to impact specific areas. I placed the headband toward the back of my head, with two coils on the parietal lobe (right and left) and two coils on the temporal lobes (right and left). I ran a thirteen-minute session with frequencies at 10 Hz and 4 Hz. Within a few minutes of beginning the session, I started feeling uncomfortable in my body. I felt a bit lightheaded and noticed my meditation heading into a deeper state that is often accompanied by a spasm or twitch in my left arm. Nothing particularly interesting happened during the meditation. It occurred to me, afterward, that perhaps I had the intensity too high on the device, causing the slight discomfort. I felt a little "spacey" after the session and went to bed. My sleep was more unsettled than usual, and I had a dream about finding a large, beautiful feather (I don't normally remember my dreams). I don't recall the specific circumstances in the dream sequence, but I do remember feeling that this feather was significant and a sign.

During my meditation the next morning, I felt different. I felt tapped in in a way that has a distinct feeling for me. I don't know how to describe this feeling, but it comes with an awareness that I am more intuitively engaged. That morning I felt attuned to the energy within my body and received a spontaneous understanding that raising my chin slightly during my meditation could help me tune in to "higher energies." As I did this, I noticed that the tone of my meditation shifted, and I was aware of energy in and around my head and directed toward the center of my brain. I spontaneously began thinking of people in my life and seeing a color associated with them. My understanding was that the color I was seeing was a frequency they needed for balance, like an individualized healing frequency based on whatever they were struggling with at the time. After my meditation I felt drawn to pull a few tarot cards. This is not something I do on a regular basis. As I looked up the interpretation of the cards in my favorite tarot book, all three of them mentioned intuitive sensing, a receptive state of being, and "listening to my inner teacher." Much of this morning meditation experience was atypical for me and the only thing different was the rTMS session the night before. This suggested that the effects of the session were delayed.

Later this same morning, I went to my favorite coffee shop to see Julie Rost. She was the owner of the café and a psychic medium. She had been the first person that came to mind during my morning meditation. Knowing that she was open to this kind of conversation, I told her about my meditation and that I was "seeing" that she needed the color blue for balance or healing. As I said that I looked at her and she had on blue pants and blue in her shirt. I commented on this, and she affirmed that we often choose colors to wear that we need. She asked me if I had any sense about the meaning of the blue. I told her that my initial reaction was related to Chinese medicine and the five elements. Blue is associated with the kidney system and the emotion of fear. As soon as I said this, Julie's face flushed. I hoped I hadn't said something wrong. She told me that she had recently been waking from sleep with a fear response related to some work-related concerns.

Despite my discomfort the night before, I started to think maybe I was on to something here, but that it needed a bit of fine tuning. I wondered what might happen if I did the same process focusing on the right parietal and temporal lobes, rather than both hemispheres. If I were to run the program at a slower frequency or adjust the power a bit, would this prevent the discomfort? What might happen if I were to try this technique on someone with more developed psychic abilities?

Julie Rost

Having already received a psychic/tarot reading with Julie, I had a good sense of her psychic abilities. I wanted to see what her reaction would be to using the rTMS stim technology and if it might add anything to her experience.

We started with a baseline EEG—just sitting there. Then I recorded another EEG while Julie tuned in. The direction of this session was left ambiguous and served as another baseline. What was it like when she tuned in in a typical way?

She began with her eyes open and gave a play-by-play of her process. She indicated that she typically begins one of her sessions by "tuning her chakras." She described beginning with her root chakra and moving up, scanning for any stale or toxic energy, and then using her directed attention/intention to spin the chakras until they are shining brightly, sloughing off any unneeded energy. This process took about two minutes, after which she imagined pouring cold water through her crown chakra to "rinse out the dust and debris," sending any excess energy into the Earth for transmutation. She then reported stepping into a room to meet her guides. She indicated that she works with thirteen spiritual helpers on the astral plane; these guides provide direction and assistance in her work. Julie described interacting with a couple of the helpers, including a Tibetan monk and Archangel Raphael. At this point, Julie closed her eyes and reported the impression of a green healing light being used to help heal a friend's shoulders and spine. She noted that her hands felt warm during this type of healing and then become cold when the healing was complete. Julie then asked her guides if there was anything they wanted her to know. As she received the information from them, she stated it out loud for my benefit. She reported hearing that inflammation can be treated through diet, by examining the health of the gut. When we eat inflammatory foods, it is similar to retraumatizing the body. We cannot heal our trauma until our gut health is in proper alignment. She reported seeing red "warbly" lines that were some kind of sign in her mind's eye of inflammation. She indicated that this sign would help her in the future. If someone comes to her for guidance and she sees these lines, like heat coming off hot pavement, she will know that it is related to inflammation. At this point, I asked if this was a good time to shift gears. Julie indicated that it was not a problem for her. She can "step out" of this psychic space and then easily "step back in."

In preparation for the next recording, I added the rTMS coils. For this experiment I was using a system designed by BrainMaster Technologies that allows you to use up to four separate electromagnetic coils. Each coil is enclosed in a small box and controlled by computer software. I decided to use two coils placed on the right parietal and temporal regions, following Dr. Powell's suggestion. Based on the research, it seemed clear that pulsing the electromagnetic wave at a slow frequency was likely to decrease activation in the area exposed. I decided to set the system to 3 Hz, in the delta range of frequencies, which is often associated with deep sleep. I started to record the EEG, hit go on the rTMS program, and waited to see what would happen.

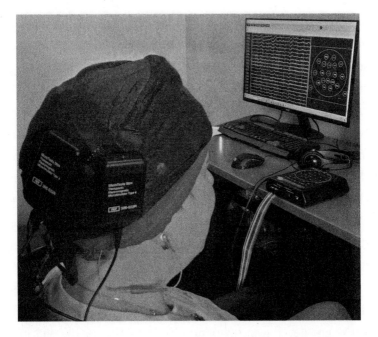

Julie reported that she was calling upon her spiritual support team and asked any of them to come forward if they had a message for her. She indicated that she saw her deceased grandmother, Stella. She reported that her grandmother was Polish, and Julie could smell specific Polish foods that her grandmother used to cook. Julie indicated that her grandmother wanted her to get excited about cabbage season coming soon. She said that she had never met her grandmother, but knew she was gifted with psychic abilities. When this interaction concluded, there was a pause of about twenty-five seconds when Julie did not say anything. When she started talking again, she said, "I don't know how to describe what I am seeing." She attempted to explain what was happening in her inner visions by stating that she was seeing a bunch of crop circle symbols. She understood that these were "keys" of some kind, recognizing that these symbols were connected to many other systems, including the chakra system and sacred geometry. She indicated that mushrooms were somehow also connected to this symbolic language. She said that it is not just the forest that understands the language of fungi, that this is a more universal language.

At about the five-minute mark of this session, Julie said, "Wow, wow! You should see what I'm seeing." Julie went on to describe visions of Mayan and Egyptian cultures and the understanding that the same symbols were used in both. She said that each civilization builds on the foundation of other civilizations, even if they occurred in very different parts of the world. Somehow, they were connected and advanced in relation to each other. As the session continued, Julie said, "What the fuck? What the fuck is this pulsing?" (referring to the rTMS), at which point she began laughing. Julie went on, continuing to translate the download she was receiving. She stated that she has never seen this kind of information before and indicated that we (as the human race) are being asked to "level up" into a fifth-dimension way of understanding. Julie stated that "believing" is a fourth dimension construct, while "knowing" is associated with the fifth dimension. After the fifth dimension everything just *IS*. You don't have to say you believe or know something because it simply *IS*.

In the last part of this session, Julie stated that psychedelics were given to society to nudge us out of a dense vibrational state. Julie then described "seeing" the base of a huge tree that had an opening, as if it had been struck by lightning. Inside this opening was the entire Universe. Julie described the scene as "very psychedelic." At this point, we wrapped it up. The entire session was under ten minutes.

When the session was over and I asked Julie if this session was any different than her typical sessions, she responded with an emphatic "Yes!" She told me that the amount of information coming through was nearly overwhelming and felt a bit psychedelic. In the session with the rTMS, she felt that she was continually being shown, in various ways, that everything is connected. That we are all One. Julie stated that her typical sessions are much more "tame."

It is interesting that the theme of the rTMS session seemed to be connection and unity. Remember, the part of the brain we were "deactivating" is related to boundaries between self and others. In fact, you might recall from Chapter 1, that disrupted activity in this region caused by head injury often results in increased empathy and spirituality.

When I glanced at the EEG recording from the session, I was initially disappointed. There were giant waves all over the screen that couldn't possibly be real. My best guess was that the rTMS was somehow causing interference with the EEG. I had considered this possibility before beginning this experiment, but remembered seeing published research studies in which they were recording EEG activity during a rTMS session. If this was done in a published study, you would assume it was a nonissue. I figured I would have to throw out all of that EEG data and just see what I could use from the experiment from the other readings Julie did while hooked up.

Being tenacious and having a sense that there may be more to the story than rTMS interference, I dug into the EEG recording a bit further. When I scrolled to the beginning of the session, these patterns were not present. In fact, they didn't show up until about the six-minute mark of the session. After the six-minute mark, it was pretty much continuous until we stopped the session. Julie had not moved her body or done anything else that might cause this change. I was continually monitoring the connection of all the electrodes which all looked fine. So, what was causing this presumed "malfunction" during the last third of the session? It wasn't caused by Julie doing anything differently, and it didn't appear to be caused by the rTMS. If it was, we should have seen the same odd pattern throughout the entire recording. The only remaining explanation that made any sense was that this represented a shift in Julie's brain.

I reduced the number of EEG channels I was observing and increased the scale on the screen. This allowed me to get a bit more clarity about what I was seeing. With this magnification it became clear that the big changes were all happening in the right hemisphere and were particularly focused around the area of the right parietal lobe. This made sense as this is where the coils were placed. The other area involved was the right frontal lobe, which has a lot of connections to the RPL. The image below shows selected EEG sensor data from an earlier section of Julie's session. This is a fairly typical looking EEG. The first two lines represent left and right frontal channels (F7, F8); lines three and four represent left and right temporal channels (T5, T6), and lines five and six represent left and right parietal channels (P3, P4). In the top image, there are two large excursions in the first two lines. This is caused by eye blinking and shows the impact of muscle and movement on the EEG.

EEG Sample Before rTMS

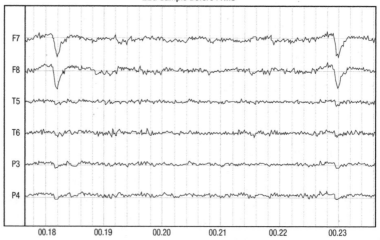

The next sample (below) comes after the six-minute mark. You can still see a few eye blinks in the frontal channels, but you will also notice that all of the right-sided channels (F8, T6, and P4) look much different than their left-sided counterparts. This pattern looks like what we see with high-powered delta brain waves.

EEG Sample During rTMS

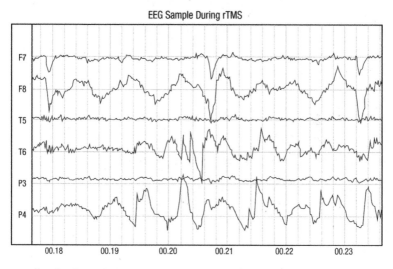

Clearly the rTMS was causing a significant dysregulation in the right hemisphere, but why did it take six minutes to show up? Apparently, the brain requires a certain level of consistent information in order to shift gears. In this case, six minutes seemed to do the trick. Persinger's God helmet sessions would usually last about thirty minutes. A quick review of other stimulation technologies and their recommended programs show that most manufacturers of these devices recommend between ten and thirty minutes to get a significant shift.

The evidence that rTMS could be used to help instigate certain paranormal experiences or enhanced intuition seemed clear. Persinger's research and my own mini experiments with Julie and a few other willing participants convinced me that I was on the right track, and that the right parietal (and temporal) regions of the brains were the most effective and appropriate targets for this approach. I also wondered if other brain entrainment technologies could have similar effects.

Audio-Visual Entrainment

Audio-Visual Entrainment (AVE) refers to technologies that involve stimulating the brain with light and sound, rather than with pulsed electromagnetic frequencies. A typical AVE device consists of specialized glasses with lights built into the eyeset and headphones connected to either a controller box or software on a computer. The user can choose from a wide range of different frequencies that are transmitted to the person by turning the signal (light and sound) on and off at the desired rate. Like rTMS, you can use AVE systems to encourage the brain toward specific states of activity (or inactivity). For example, if you were interested in enhancing theta brain waves, you could choose a 7 Hz pulse rate. The lights would flicker on and off seven times each second, synchronized with a tone in the headphones. The brain attempts to match this frequency through something called an *evoked response*, resulting in an increase of the stimulated brain wave pattern.* In Chapter 3 we discussed using AVE technology for the Ganzfeld effect, which is a static approach— simply leaving the lights on the entire time. I was interested in also using this tech to encourage specific brain waves that appear to be related to psi abilities.

A caveat: The process is more complicated than the brain simply following the desired frequency. In fact, in some instances the frequency stimulated by AVE inhibits that specific activity in the brain. A fuller explanation is beyond the scope of this book. Please see the Resources section for additional information.

Psychic Exploration Class

While working on this book, I began teaching a four-week online Technology Assisted Psychic Exploration class. As part of the registration, all participants received a pair of AVE glasses. Each week, I would present information related to the brain science behind psi abilities, and then we would experiment with an AVE protocol designed to facilitate the associated brain-wave patterns. Participants would then work with that protocol for the next week until I introduced a new approach.

Even though I have more than twenty years of teaching experience, I was a little bit nervous beginning this class. While I had tested these protocols in my office, I didn't know if this approach could work with larger populations and in group settings. The whole thing was a grand exploration. Now, after teaching this course multiple times, I can comfortably say it appears to work, sometimes with quite dramatic results.

Student Experiences

While discussing their experiences with a theta brain wave protocol, one student in the class described connecting with their Higher Self.

"My Higher Self came to me in the form of a geometric shape that was red, but it's not like one I had seen before, like a really weird M. It mostly stayed in the form of a red shape, and it was masculine, which was unusual for me. It was basically just telling me that a lot of really great things are coming, and I have to trust in the process and that I am never actually alone—there is a part of me that I am retrieving. Then it laid down into me in the chair and I felt this sense of something else that is totally me, sitting with me. It was a really pleasant experience."

Another student described how a recent session seemed to have activated some kind of deep connection to the Earth that has remained with her outside of the AVE session. "I couldn't get to sleep the other night and did a delta session and it felt fabulous. My physiology loved this frequency. I stayed there for a while and eventually fell asleep. A few hours later I woke up experiencing a consciousness deep inside the Earth. It is communicating through what feels like an ancestral voice. I can feel the inside of the Earth's consciousness and all of the ley lines and grids. There is very vivid imagery and a felt sense. I experience it in all of the sensory modalities—what it's like to be the grids, the meridians, the soil, the bacteria of the Earth from different periods of existence. I can feel this sense of other extinction periods and the Earth really talking about 'Do you have any idea what it's like for us to have this much population? Do you have any idea what it's like for us to have this much concrete; to not be able to breathe; our qi being depleted?' It has been so real and so deep, and I am experiencing this during normal life. It has been so profound. It is like this voice that wants to be heard. It's so powerful, I can actually feel rumblings through the Earth as if it is saying, 'This won't be tolerated much longer, and we wanted you to know.'"

Of course, these examples are not fully representative. In some cases, the experiences are relatively mundane. Another participant in the same class provided this report of their experience:

"I did a theta protocol before our last class. I always feel stoned afterward. I look affected. It feels really good; I love it. I am doing that [theta protocol] and trying to balance it with an alpha protocol afterwards. It's been wonderful. Nothing psychic, but very visual, very bodily. I want more cool stuff, but I'm just getting out of the way."

This type of experience is fairly common. It sounds pleasant but doesn't seem to be leading to any enhanced psi experiences.

Based on these and other reports, I have identified three themes that are important to consider when using tech for psi exploration.

Individual Differences. Not surprisingly, not everyone has the same response to these sessions. While some participants report powerful and personally meaningful experiences in their AVE sessions, others state that it was "nice," or "fine." In some cases, students might even dislike a session or find it unpleasant. For this reason, it is important to experiment with a variety of protocols and settings. Sometimes slightly modifying the frequency can significantly change the experience. In other instances, it is necessary to change the color or intensity of the lights, or other aspects of the program to maximize the benefit. When students spend time learning how to use the system and tracking their experiences, nearly everyone is able to find at least one approach that works for them.

Psychics Versus Nonpsychics. In general, it appears that the AVE and rTMS techniques have stronger effects on those with a background in psi-related activities. In the personal accounts described above, the first two participants have a relatively strong background in psi-related practice, while the third person does not. This trend has been regularly seen across groups and suggests that stimulation technology amplifies preexisting abilities. You may recall that we saw a similar process in Chapter 3, when participants were using the Ganzfeld technique with white noise. This observation is consistent with the finding that the technology makes it easier to tap in. If you already know how to extend or expand your consciousness in psi-related ways, the technology seems to give it a boost. If you do not know how to tap or tune in, it may be necessary to explore more traditional psychic development practices alongside the use of technology. Simply zapping your brain may not be enough.

Delayed Response. The report from the second student above as well as my own experience, helps demonstrate another theme.

In some instances, the impact of the AVE or rTMS session isn't experienced until after the session has ended. This is highlighted in another comment from a class participant: "I got much more of an effect from the session after the session was over than when I was experiencing the flickering lights directly. In terms of getting back to an old feeling of heightened intuition, I felt it was more useful to run the session and then stay quiet, leaving the lights off and using the state that came after the lights were off."

This person happened to have a background in neurobiology and offered a possible explanation for this experience: "My model is that when the brain is doing any kind of complex activity, lots of different brain areas are being activated in very complex ways. It seems to me that when you are putting yourself in front of a flashing light that is driving the brain, there is a possibility that it [the AVE] is not letting the brain go to those [complex] states, but rather biasing sensitive centers in the brain toward a certain frequency, making brain centers more able to do what they might be likely to do. But if it is running constantly, it is not giving other processes a chance to knit that activity together."

While it seems clear that brain stimulation technologies can have a significant impact on psi-related experiences and intuition, the mechanism of action is poorly understood. Suffice it to say, the brain is highly complex and there are myriad individual differences that may influence how each person responds to specific stimulation devices and frequencies. The only way to know for sure how a particular technique will impact your abilities is through self-experimentation.

Try It Yourself

Since this chapter deals specifically with using brain stimulation techniques, all of the "try it yourself" approaches involve some kind of technology. The good news is that there are some effective, easy to use, and relatively cost-efficient options available. Below, I outline two of the consumer-grade devices available that we recommend and use in our classes.

NeoRhythm

The NeoRhythm is a low power rTMS device and what I used in my self-experiments described above. It is designed in the form of a headband that communicates to a phone app through Bluetooth. You can find out more about this device by checking our website at: www.psychicmindscience.com. When working with the Neo-Rhythm or similar devices, there are a few considerations that may be helpful.

Power. Anything that uses electricity also has a magnetic field. The strength of those magnetic fields are measured in *microteslas*, abbreviated μT. While exposure to prolonged or high-power magnetic fields can be harmful, there is no evidence that exposure to low-power magnetic fields is problematic. If anything, research suggests that low-power magnetic fields, when used strategically, can enhance healing. The amount of power used in the NeoRhythm is relatively small (between .25 and 2.5 μT). For perspective, the field strength of most appliances in your home ranges from .02 to 7 μT. Even though we are dealing with relatively low-power devices, they can still influence the brain and states of consciousness. In addition, different individuals seem to have varied levels of sensitivity to this form of neurostimulation. For this reason, I generally recommend starting with a power setting around .5 or 1.0 μT and assessing the effect. If you feel dizzy, lightheaded, or in any other way uncomfortable, try shifting to a lower power. If you don't notice anything, try turning it up a bit.

Placement. If you want to disrupt functioning in the right parietal lobe, you will want the coils to be on the right parietal and temporal lobes. You can do this by wearing the front of the NeoRhythm headband on the back of the head, just where the head rounds down from the top. The arms of the headband can extend down toward the ears. If you are interested in stimulating gamma activity in the occipital lobes as discussed in Chapter 4, you would place the front of the headband at the back of the head so that the arms of the headband extend straight across the head, just above the ears.

Frequencies. This will take a bit of experimentation. In general, if you want to disrupt or shut down an area, like the RPL, you will choose slower frequencies, often between 1 and 3 Hz. If you want to stimulate an area, you will typically choose frequencies in the gamma range, 35 to 80 Hz.

Length of Session. Typical sessions range from ten to thirty minutes. You may need to experiment a bit to see what length of time provides the desired impact. Most people seem to notice something between ten and twenty minutes. If you don't notice anything after thirty or forty minutes, you may need to adjust the frequency, placement, or intensity.

Contraindications. This information is taken from the Neo-Rhythm website, which states that the device is generally safe for use. However, people with certain medical conditions should *NOT* work with this type of technology. These include:

- People with pacemakers or other electronic implants, cochlear implants, or mechanical heart valves
- Women with nonMRI-safe intrauterine devices (IUDs)
- Persons with a seizure disorder
- People with active bleeding
- Organ transplant patients
- Children under the age of twelve

While it is not specifically contraindicated by the manufacturer, I also recommend caution for anyone managing significant mental health concerns, including schizophrenia, bipolar disorder, or dissociative disorders. If you have any medical or mental health condition or any concern regarding the use of this type of technology, I recommend consulting your healthcare professional before using NeoRhythm or similar devices.

Spectrum USB Audio-Visual Entrainment Glasses

These glasses are produced by a company called Mind Alive and were described in the Chapter 3 try-it-yourself section. The glasses connect via a USB plug to your computer where they are controlled by an easy-to-use program that can be downloaded from the Microsoft app store. Unfortunately, the software is not compatible with

Apple devices. Consequently, if you use a Mac computer, your best option is to purchase one of the stand-alone devices, such as the DAVID Delight.

The DAVID Delight and upgraded versions of this device (DAVID Delight Plus and DAVID Delight Pro) use similar glasses but include a small box that controls the programming. These devices cost a bit more and limit your choice of protocols to those that are predesigned by the company. If you are going to invest in this type of system, we recommend either the Plus or Pro models as they offer more options. All of these models can be purchased through the Psychic Mind Science website: www.psychicmindscience.com.

Protocols. Regardless of which AVE system you are using, protocols with a target frequency in the theta (4-8 Hz) and gamma (35-45 Hz) ranges appear to be the most helpful in tuning the brain to psi. However, everyone is different. Some people prefer slower protocols (delta) or protocols in the alpha range. We also offer specialized protocols that shift between a range of frequencies during the course of the session.

Other tips:

- Drink plenty of water before and after an AVE session.
- Consider using music or a guided meditation along with the light stimulation, particularly if the tones are unpleasant.
- If you are new to AVE, start with the volume of the audio and brightness of the lights at a lower intensity.
- If you are new to AVE, try using an alpha session for the first time or two.
- Keep notes regarding your subjective experience of each protocol you try to help identify which approach(es) work best for you.

- Trust your instincts. If you like a session but can't identify specific changes, assume that something is happening, but hasn't manifested yet.

- Remember that effects may be delayed, so watch for changes later in the day after a session or even the following day.

- Have fun!

CHAPTER SEVEN

PSYCHOMANTEUM

GREG WAS A MENTOR, spiritual guide, and my first qigong teacher. He had been sick for some time but had been very private about it. He only told a handful of people what was happening and had mostly secluded himself. We knew he was battling cancer, but his death was not expected. Greg was a fighter and skilled in various methods of healing. He died of complications from pneumonia. It happened suddenly and was a shock to everyone, including his wife and kids.

One of my friends and a member of our healing group called to tell me the sad news. We both felt disbelief and regret that we had not reached out to him more or offered more support. The truth is, Greg had some kind of stroke several years earlier that left his personality a bit more eccentric than before. He was harder to connect with, and it felt like he was not as sensitive to environmental cues. We all lost touch in the several years before his death.

The loss of someone often makes us realize how important they are to us. What an unfortunate and sad set of circumstances.

Two months after his death, Annette, Greg's wife, contacted all who had been impacted by Greg through his healing and teaching. She invited us to come to the house and select something from his healing room to take with us. This was a unique and powerful experience. I arrived with my first *Taiji* (tai chi) teacher. I had introduced him to Greg many years before, and they had shared many similar interests and practices. We arrived and greeted Annette, who was clearly still heavily in the grieving process. She seemed glad to see us and gave us both very strong hugs. I again had a huge wave of guilt pass over me, wishing I had been more present and helpful for her during this difficult transition. She eventually took us upstairs to a sitting room where there were five or six other students/friends of his. It was a small room with a loveseat, television, and bookshelf. Based on the books, various shamanic tools and prayer beads, it looked as if this was Greg's reading room. We were all seated in a circle, telling stories about how we met Greg, things he had said and done, detailing the quirks of his personality. Annette told us to take whatever books we wanted. It is interesting how you can tell something about a person based on their books. His library included a variety of classics and obscure texts on Taiji and qigong, as well as various forms of healing and spirituality. I ended up with an armful of books to take home—one of which was mine that Greg had borrowed ten years earlier.

We took turns spending time in Greg's healing room, which was adjacent to the reading room. This was a tiny room with a chair and shelves, altars, and small tables everywhere. Every surface was completely covered with spiritual and healing objects, many of which Greg had made himself. When it was my turn and I entered the room, I immediately felt a very strong energy. I am not generally super-sensitive to energy, but this was palpable. My heart was racing, and I felt altered. I sat in Greg's chair and soaked it all in. There were stones, crystals, fetishes, statues, jewelry, and healing tools everywhere you looked. It was overwhelming. I scanned the room looking for any objects that caught my eye or were meaningful to me; things I might take with me as a memory of Greg, or items I might use in my own spiritual or healing practice. I was first drawn to a white statue of Kwan (sometimes spelled Kuan or Quan) Yin, the Buddhist Goddess of Compassion. I also chose a brass bracelet, a dream catcher that Greg had made, a couple of gems, and a bear fetish (Greg was known by several Native American names, many of which included Bear). The energy in the room was so magnetic, I literally had to pull myself out of the space. I could have easily sat there for an hour or more but was aware that others had not been in the room yet.

Interestingly, I was not the only one with this experience. Nearly every person who entered the room would come out saying something like, "Woah! That was intense."

The rest of that day I had that strong, tingly sensation in my chest that had started in Greg's room. I felt happy and almost joyous despite the fact that this was a sad occasion. I had a feeling that Greg's energy was with me—with all of us. I could feel his presence, although I didn't know how to define it. I think I was not sad because I knew he was there. Not only did I recognize his energy, but I felt like somehow this whole process was a lesson for all of us, a teaching tool—a way to reach all of us at the same time with powerful healing and understandings. I did not know what any of these healings or lessons were, but I sensed they were happening for each of us. How do you describe that clear sense of knowing that is not based on anything tangible? There is no evidence for it, you just know it to be true.

I put the power objects from Greg's room on my altar; they seemed to fit nicely. Throughout the next several days I continued to have this very clear sense of Greg and his energy. Again, it was as if I could feel his presence. I also felt like he was reminding me to return to my healing work. This was something I began with Greg and then neglected for many years. Being in his space was a very strong reminder of that work, of how powerful it was, and how much I felt drawn to this way of helping others.

Even though Greg had left his body, it felt as if he was still there. It felt like he was with me (or near me) and still teaching lessons, as if he had messages for me and an invitation to connect with him in a different, more ethereal manner. Of course, the scientist part of me wanted to know, *Is this real?* Is Greg's discarnate soul somehow trying to reach out to me and I am fumbling around trying to connect with it in any way I can, or is this all simply some part of my psyche playing out a wishful thinking fantasy?

From a subjective, first-person perspective, I'm not sure it matters what was really happening. My experience of Greg and his continuing presence was real to me. I can't prove that he was still there. I only had my feelings. In this way, the afterlife seems to have a lot in common with spirituality and religion. Much of it is based on faith and subjective experience or belief.

From a scientific viewpoint, you might argue that my brain was simply recalling elements of Greg from my history with him. All of the previous interactions with Greg were blending into some sort of a schema that my mind created, a template of Greg. Now that he was gone, my mind was replaying and reimagining these elements. Perhaps it was nothing more than a memory or a fantasy, a way to ease the suffering of his loss.

This raises several interesting questions about mediumship. (1)

Is it possible for incarnate humans to somehow communicate with or receive information about those who have died? (2) If so, is that information coming from the deceased, or is it information being picked up from the bereaved? (3) Does it matter?

The Psychomanteum

Only a few weeks after the experience with Greg, I was scheduled to spend a few days in Long Island, New York, conducting some more EEG experiments with psychics and mediums. I was staying at the home of Bob and Phran Ginsberg who run the Forever Family Foundation. They told me they had constructed a *psychomanteum* in their basement that was built to the identical specifications used by Arthur Hastings in his research (2002).

When I first heard they had a psychomanteum, I had to look it up—I had no idea what this thing was. Basically, it is a small room, covered entirely in black. In this case, it was constructed out of sturdy PVC pipe with thick black cloth hung on all sides and a black rug on the ground. Inside this black box is a black office chair that can recline. Accompanying the chair was a small black footrest. This allowed the person inside to sit in the chair, lean back, and put their feet up. Directly across from the chair, hanging toward the top of the wall was a mirror. The mirror was positioned in such a way that the person sitting in the chair could not see themselves. There was only one small night light on the back floor of this compartment. All of the lights outside the psychomanteum were turned off. The instructions given to me were to relax in the chair and simply gaze at the mirror, allowing any experiences to naturally emerge. footnote: In formal psychomanteum research studies, the instructions are a bit more elaborate and involve identifying a particular deceased person that the sitter wishes to contact. The sitter is encouraged to talk about the deceased in great detail and then clear the mind of everything except the deceased individual (Moody and Perry, 1994).

Because I tend to experiment on myself and was interested, I found an opportunity during the early morning to hook myself up with an EEG cap and place myself in the small, dark chamber. I kept the computer outside of the contraption to eliminate any additional light.

Almost immediately, I had this very strange bodily sensation that lasted for nearly the entire forty-five-minute session. It felt as though the atoms in my body were buzzing, like they were kicking into a higher rate of vibration. This was a very unusual sensation and caused me a little anxiety at the beginning. Not because it was unpleasant, but because I wanted to move my body in response to the energy and because the way these types of sensations are typically experienced is related to nervousness or panic. There was also something about the situation that felt "creepy," for lack of a better term. It was not that anything bad or scary was happening, it was more of a feeling associated with the unknown. It reminded me of times when I was a kid and would play with a Ouija board and felt that I was touching something beyond this world.

As the session progressed, I found my mind entering a kind of trance state. My body felt energized and fast, but my mind felt quiet. I started to see images. It was odd, because it was the kind of imagery you might have when you are sitting with your eyes closed, but in this case my eyes were open, gazing at the mirror. At one point I saw a few faces. It was not scary, but it was unexpected. The images were vague, and I could not necessarily make out distinct features to tell who they were. While still in the experience, I focused on one face that had a distinct shape. As soon as I did that, I recognized the face as my Great-Grandpa Berg. It was not that I saw his face clearly,

I just knew that this is who it was. This sense of knowing was strong and clear. It also did not feel like I was trying to create meaning—it was just obvious. Nothing else happened with these faces. They were there and then they were not. Throughout the session I would see things in my mind's eye, and often it was difficult to hold on to the image in a conscious way. It was almost like a dream state where you wake up and then must remember what just happened. Some of the images that moved through my mind included a horse's head and a large ball of light. There was also one point during the session where I mentally "asked" if any entities present could identify themselves. Almost as soon as I sent out the question to the spirit world, I forgot about it. (Remember I was in a fairly heavy trance state, making it hard to hold on to normal mental functions.) A few minutes later, I very clearly heard the name "Lucy." This was odd for me as I do not know anyone named Lucy and have no way to connect this information to anything, however, the name was crystal clear.

Probably the most impactful moment of the session occurred near the beginning. I was not thinking or intending anything, and I heard in my mind someone say, "Just because I am no longer with you, doesn't mean I stopped loving you." It was a female voice that I heard in my head, but again it did not seem to be coming from me. It was a very comforting thing to hear and felt like it was more of a general message for everyone rather than something specific for me. It gave me the feeling that the deceased are somehow still connected to us when they die. They are looking over us and working with us in ways we don't perceive. It was an understanding that love is what it is all about. I actually did not share this experience with anyone after I left the psychomanteum. I felt a little embarrassed by it, like it sounded contrived in some way, so I just kept it to myself.

As I left the psychomanteum I felt dazed. I did not want to talk and was in an altered state. It took me a solid hour of walking around, eating some food, and processing my experience (except for the love message) with the rest of the group before I felt more normal. As I think about it, I am not sure I felt "normal" at all for the rest of the day.

After my session, but before I shared my experiences with the larger group, my wife, Erika, decided she wanted to try it as well. After her forty-five minutes in the dark, she emerged and looked a little stunned. She went straight outside and put her hands on the Earth. After some time, she reentered the house, ate some food, and started to share.

As we took turns processing our experience, it was interesting how many similarities there were. Erika also saw faces and lights. She reported that this scared her a little bit, but she was able to breathe through it. However, at one point she clearly felt the presence of "someone" beside her and that totally freaked her out. Somehow, she managed to stay in the room until her time was up, but just talking about it brought tears to her eyes. She was obviously shaken by the experience.

Because this was the first exposure either of us had to a psychomanteum, we had no idea what to expect. It wasn't until later that I found research showing that our experiences were pretty typical. Moody, the originator of this technique, found that 50 percent of the people entering the psychomanteum reported experiencing some level of communication with a deceased loved one (Moody, 1992; Moody and Perry, 1993). In another study, 48 percent of the participants reported contact with the deceased, while 70 percent reported some level of "anomalous experience," including visual hallucinations, a feeling of a "presence" in the room, OBEs, sounds, voices, smells, and time distortion (Hastings et al., 2002).

I also discovered that the original researchers working with the psychomanteum had a very strict protocol they used for people coming into and out of the experience; in fact, they would sometimes reserve three to six hours of time for a single session. The client would spend time before the session focusing on a particular loved one they wished to contact (setting an intention), telling stories about them, and maybe even bringing in objects connected to that person. After the counseling session and additional orientation, they would enter the psychomanteum with a sitter available just outside the curtain. The sitter was there to assist if there were any problems or concerns. As I thought about my experience, it seemed obvious that just having someone present may have been comforting during times that were more challenging. At the end of the session, the person in the psychomanteum could choose to stay for an additional fifteen minutes or come out. They would often be encouraged to crawl out, staying low to the ground. This may have been helpful as a grounding technique but also to prevent falling, as the experience can be very disorienting and even dissociating. The client is given time to sit quietly and write down reflections from their experience and then as much additional time as necessary to discuss any of their thoughts or reactions. After my experience in the psychomanteum, I was left feeling that somehow, I had cracked open a doorway into some other realm and peeked inside. I think this is partly why I felt so unsettled afterward, and why I initially had the sense of it feeling creepy. In fact, days after the experience I continued to have this vague sense that I was "opened" up a little bit. I felt a little vulnerable.

The Brain on Psychomanteum

When I had "recovered" from my experience, I analyzed the EEG data, comparing my baseline recording to the psychomanteum data. The results were striking. Compared to baseline, there was a significant decrease of slow brain wave activity, especially in the theta and alpha brain wave bands, and especially toward the back of the brain at the hub of the default mode network (represented by the lighter shades).

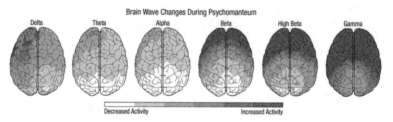

Brain Wave Changes During Psychomanteum

This was somewhat surprising to me as I had mapped my brain dozens of times, always showing an excessive amount of alpha brain wave activity. In fact, there was one point early in my career when I was trying to reduce my alpha through neurofeedback, thinking that too much alpha may be a problem. After forty sessions, there was no movement. To see such a striking shift within a relatively short time was dramatic and suggests a significant change in the way my brain was tuned to the environment.

Alpha Brain Waves

In general, you can think of alpha brain waves as relating to an internal versus an external point of attention. When there is an increase of alpha, a person is usually focused internally—caught up in thoughts, daydreams, or even a silent meditative state. When alpha decreases, the attention is external, as might be seen when you have your eyes open and are focused on something in the environment.

It is curious to see such a dramatic decrease of alpha (external attention), given that I was sitting in the dark with nothing to look at. Perhaps the very slight light in the box and the focus on the mirror (and any strange visuals in the room) was enough to cause this brain shift. In the absence of any normal visual information, perhaps my brain was working extra hard to make sense of anything that could be visually perceived. It is also important to consider the role that alpha plays in the default mode network (DMN), the area where this decrease was most prominent.

The DMN and Alpha

The DMN, which is discussed in more detail in Chapter 8, is largely involved with creating a consistent narrative about the self. It is the neural mechanism that coordinates all of our thoughts, memories, and plans into a coherent whole, attempting to make sense of the world and our place in it. It is an identity-making machine. Alpha is the dominant brain rhythm in this network, with the job of keeping this system organized and stable. If you disrupt the normal functioning of the DMN by decreasing alpha, it opens the brain to consider new perspectives and become more open to additional information. Not surprisingly, this pattern of decreased alpha in the DMN is the most common brain pattern observed during psychedelic states, disrupting the status quo, and facilitating cognitive flexibility.

Frontal Lobe Activation

The other highly significant finding from my self-study was the dramatic increase of fast brain waves in the frontal lobes. This extra activation may seem counterintuitive. In fact, you might predict that the frontal lobes would become quieter in the absence of sensory information. Perhaps this is precisely why the frontal lobes are hyperactive. There is nothing for it to process, so it is actively searching for information, tuning in to more subtle layers of sensory experience that require heightened activity. Interestingly, this increase of frontal lobe activation is a pattern sometimes seen with psychics and mediums. Recall that Elisa and A. J. both showed increased frontal-lobe fast activity during telepathy and spirit communication sessions (see Chapters 3 and 5).

Brain Analysis Interpretation

Taken together, the brain wave analysis seems to suggest that the environment in the psychomanteum led to an altered state of consciousness that allowed me to let go of preconceived ideas and perceptions about myself and my relationship to the spirit world. In addition, the lack of sensory input created an opportunity for my mind to focus on the black space in an active way, facilitating the perception of more subtle sensory cues such as voices and faces.

A Skeptical Perspective

I didn't enter the psychomanteum under controlled conditions, and I did not follow any of the protocols established by the early researchers in this field, yet there was something about this experience that led to a perceived contact with disembodied energies. The research evidence shows that this type of sensory deprivation experience consistently leads to some kind of "contact" experience, but how can we determine if these experiences are "real"? What role does expectation play, or does it matter?

The fact that the typical psychomanteum process involves spending an extended period of time explicitly focusing on the deceased person you want to contact before entering an altered state is seriously problematic from a scientific perspective. It could be argued that the psychomanteum process is putting the client into a hypnotic, trance state where they are more suggestible and experiencing a fabricated perceptual set based on the client's desires. One of the leading researchers in this field directly discusses this problem, describing that many of the psychomanteum experiences involve demand characteristics and suggestive aspects that are likely to influence the person's experience (Caputo et al., 2021). If you think about the process, there are numerous red flags. Let's look at it from a skeptical viewpoint.

First, the psychomanteum experience itself is designed and "marketed" as a technique to contact the dead. So, before you have even done anything, there is selection bias, and the mind is already starting to set up certain hopes and expectations. You can imagine a scenario where someone hears about the psychomanteum, is interested in trying it because they have unresolved grief related to the loss of a loved one, and starts reading other people's experiences on the Internet. Because our minds automatically seek and attend to information that confirms our beliefs, you only pay attention to the psychomanteum stories that describe meaningful communications between the person and their loved one. You repeatedly think about and imagine how your own experience might unfold. Then the day of your psychomanteum journey arrives. You talk about your loved one for a few hours with a sympathetic counselor, telling stories and sharing mementos you brought that remind you of the deceased. Then, you enter the chamber. It is almost completely black, with just enough light to see a mirror elevated above your sight line. You sit in this space for forty-five minutes, looking at the mirror with nothing else on your mind but your loved one. The environment creates an altered state of consciousness, opening the mind to increased creativity and imagination.

Under these circumstances, it should not seem odd or unusual to have some kind of paranormal experience that involves the subjective sense of spirit communication. The idea that suggestion or expectation might play an important role in the psychomanteum experience has not been lost on researchers. In one study, Terhune and Smith (2006) divided psychomanteum participants into two groups. One group received a neutral set of instructions while the other group received a suggestive set of instructions. These suggestions involved informing the participant that they may experience unusual body sensations, strange visual or auditory experiences, a felt "presence," or an OBE. Not surprisingly, the group that received the suggestions were much more likely to report unusual visual experiences and hearing voices, as well as a higher likelihood of dissociation.

The Role of Expectation and Intention

While focusing on a specific desired outcome (connecting with the deceased) may be problematic from a scientific perspective, it may be an important element in facilitating a spirit communication experience. If you are not mentally and sensorially open to perceiving subtle information, you will not see (or hear) it. Part of "tuning yourself" to the right frequency may involve cognitively and emotionally connecting to the energy of the person that has died. While we do not understand the mechanisms behind how this works, it seems to be an important consideration. In fact, in our explorations at the NeuroMeditation Institute, we have found that participants that are already attuned to psi-related phenomena are much more likely to have significant experiences than those who are casual observers of the experience. Marilyn Schlitz, PhD, a Senior Fellow at the Institute for Noetic Sciences (IONS), has indicated that it is helpful to "enhance expectations of connecting with invisible others and engaging in meaningful communication with them." She indicates that this aspect of the psychomanteum preparation invites active imagination and depth of experience (2021). Perhaps this process is akin to the important role of set and setting in other altered states experiences. We know from decades of clinical and research work with psychedelics that a person's mind(set) and the environment (setting) make a huge difference in determining the person's experience.

Evidence

While it is easy to explain away some of the psychomanteum experiences as due to suggestion or expectation, it is difficult to shake the felt sense and emotional impact of the experience. I wondered if there was any evidence to support the idea that these experiences were real and not just figments of my imagination. Considering this question, a study conducted by Christine Simmonds-Moore, PhD, a professor at the University of West Georgia, combined the psychomanteum with ghost-hunting tools. Dr. Simmonds-Moore recruited a group of *synesthetes* and nonsynesthetes to participate in a psychomanteum session. Synesthetes are people who experience sensory information in mixed ways; for example, they might "hear" colors, or "smell" sounds. Because of this ability, they tend to be more sensitive to paranormal activity. After the psychomanteum experience, each participant was asked a series of questions about their experience, focusing on any sensory perceptions that may have occurred (seeing a figure, hearing a voice). In addition, this study included a variety of physical field-detection devices commonly used in ghost hunting, including a random event generator (REG), geomagnetic field detectors, and infrared cameras. When the final data was analyzed, they found anomalous readings on the various instruments that corresponded with times the synesthetes reported a perception that an "invisible other" was taking up physical space in the room (2021).

A somewhat similar study conducted by Dean Radin (1996) measured a variety of physiological states (e.g., brain waves, heart rate, skin temperature) as well as aspects of the physical environment (e.g., infrared light, temperature, electric, and magnetic fields) during the psychomanteum experiences of seven volunteers. All of the volunteers reported experiencing some kind of "mild apparition." When the data was examined, there were significant correlations between the environmental and physiological data. Framed another way, when energy in the room changed significantly, so did the energy of the person. Because the room was protected from all outside influences through specialized shielding, the only source of energetic shifts had to come from within the room. Dr. Radin suggests two possible explanations for these findings. First, intense changes in the individual caused changes in the physical environment, *or* changes in the physical environment (potentially caused by noncorporeal entities) were registered in the physiology of the participants. Of course, neither of these studies prove anything, but it sure adds some confidence to the idea that something real is happening and that these experiences are not just "in our heads."

After-Death Communication and Grief Therapy

Whether these experiences and communications with the deceased in the psychomanteum are real or not, they seem to represent a powerful and underutilized healing tool. The grief commonly experienced with loss of a loved one can be devastating. In addition to the pain of loss, there is often a need for continued connection. Many approaches to grief therapy are based on the notion of "breaking bonds" with the loved one; letting go in an attempt to move on. Unfortunately, this paradigm may contradict the natural grief process and be specific to Anglo-American practices (Davies,

2004). The continued need for attachment to the deceased and the desire to remember them is often neglected in this approach. An alternative to the breaking bonds approach is the Continuing Bonds theory or CB (see Klass et al., 2014). This perspective is based on research indicating that the bereaved typically want to understand the loss (meaning-making), integrating the loss into their lives while maintaining a bond with the deceased (Neimeyer, 2014). In fact, researchers have found that higher levels of meaning-making were associated with better grief outcomes in the two years following the loss. A natural part of the CB process involves after-death communication (ADC), which can include a sense of the deceased presence, as well as sensory experiences related to the deceased (voices, smells, visions), and dream experiences of the deceased (Barbato et al., 1999). These experiences of the deceased, whether subtle or overt, often occur spontaneously or may be triggered by physical reminders of the deceased—such as my experience of Greg's presence while visiting his meditation room. Some modern approaches to grief therapy intentionally strive to induce an ADC, including working with a medium, engaging in specific meditation practices, using induced lateral eye movements (IADC; a specialized form of EMDR), and of course, working with a psychomanteum. While an examination of all of these methods is beyond the scope of this chapter, the research suggests that these methods consistently lead to a positive healing experience and reduction of grief.

After reviewing the literature on after-death communication as a method for grief therapy, Wassie (2022) reported three important considerations for this work. First, regardless of the specific method employed, the role of the facilitator seemed to be important. Not only does it seem important to have a guide that is experienced with the specific method, but also trained to provide support for those struggling with strong emotions. Second, participants who achieved a relaxed and receptive state were more likely to experience an ADC. This suggests that it may be important to incorporate relaxation training as part of the preparation phase of this work. Finally, the common use of eye movements and mirror gazing suggests that incorporating some form of visual stimuli in the experience may assist the process.

Try It Yourself

After my psychomanteum experience, I wondered if there were any easier ways to do this work. Clearly, the psychomanteum is effective but it is also a complicated setup. Fortunately, there are a couple of easy solutions. Rather than building an elaborate black box with the standardized equipment requirements, you can simply gaze deeply into a reflective surface, such as a mirror, crystal ball, body of water, or try a virtual-reality psychomanteum experience.

Mirror Gazing

Staring into a reflective surface as a means of altering consciousness and contacting the spirit world dates back to ancient Egypt and the Orient. Before mirrors, these ancient cultures would stare into standing water or oil-based surfaces. Modern approaches to this method typically involve staring into a mirror in a room with low-level illumination. In general, the setup for this experience simply involves the observer staring at their reflection and maintaining a fixed attention on their eyes or nose (Caputo, 2010). Typically, these sessions only last ten minutes or so, which also makes it much more user-friendly compared to the forty-five-minute psychomanteum experience.

In mirror-gazing sessions, it is relatively common for the observer to have a range of anomalous experiences, but they tend to be almost entirely in the visual realm. For example, it is common for observers staring into a mirror to notice lip movements—as if the reflection wants to say something, or a double apparition, which stands to one side of the observer's reflected image.

Mirror-Gazing Meditation

1. **Check Your Environment.** Because this practice requires a degree of focus, it is best to take a few moments to create a space that will facilitate a relaxed but alert state of mind. Dim the light in the room as much as possible without losing the ability to clearly see your reflection. (You can also try using a candle for illumination.) Make sure you will not be disturbed during your session and get rid of any distractions (like your phone).

2. **Mirror Setup.** Angle a mirror in front of you so that you are face-to-face with your reflection, staring into your own eyes or at your nose (you might try it both ways and see which you prefer).

3. **Mental Preparation.** It can often be helpful to spend a few minutes before the mirror-gazing meditation to quiet the mind. You might practice a few minutes of slow, relaxed breathing, focusing on extending the exhalation or a short body scan directed toward releasing any unneeded tension in the body.

4. **Set a Timer.** While this step is not absolutely necessary, it can be helpful. Sometimes time will become distorted, or you may become overly focused on *How long have I been doing this?* By setting a timer to the desired length of the session, you can just relax into the practice and trust that the timer will alert you when you are finished. It is generally recommended to begin with five minutes and then work your way up to ten minutes.

5. **Lock Eyes.** Stare directly at your own eyes (or nose). Hold this gaze as much as you can without shifting attention. Stay relaxed by breathing slowly and observe any changes in your reflection or in the space around your reflection. Continue until the timer signals the end of the session.

6. **Journal.** Spend a few minutes reflecting on your experience. What did you notice? What thoughts or feelings showed up? Were there aspects that were difficult? Are there things you would modify for your next session?

7. **Practice.** As with any skill, mirror gazing can evolve and grow with experience. Once you have the basic formula, experiment with modifications. Change the environment, your preparation, and the length of time. Have fun and follow your intuition.

Virtual Reality

In the process of writing this chapter, I connected with Dr. Marilyn Schlitz. Marilyn is one of the preeminent researchers in fields related to parapsychology. She has conducted studies on remote viewing (the ability to gain information about a distant place or person through clairvoyance or out-of-body travel), Ganzfeld, distance healing, remote staring, as well as experiences of contact with departed loved ones. For the past few years, she has been conducting research using a virtual reality psychomanteum-type experience as a method for inducing ADCs. Marilyn invited the NeuroMeditation Institute to join her most recent study, allowing me to gain firsthand experience with the power of this intervention.

If you are not familiar, virtual reality (VR) involves wearing a headset that generates realistic images, sounds, and other sensations in a 3D-immersive environment. This approach is very effective as the brain interprets the experience as real; that is, seeing is believing. The psychomanteum experience created by Dr. Schlitz's team begins at the base of a flowing creek near sunset. The presence of the flowing water is designed to induce a trance-like state as well as mimic the reflective surface aspects common to traditional psychomanteum experiences. The participant hears guided instructions, encouraging and suggesting the likelihood that they will encounter "invisible others." Eventually, the guided audio and reflective creek scene fade to be replaced with a dark background with wisps of smoke-like images appearing and disappearing in the screen. The participant is then invited to remain in this space for approximately eighteen minutes, verbalizing anything they see, hear, feel, or any information that they may receive during this session.

Of the sessions we have completed to date, 100 percent of the participants have reported significant anomalous experiences, many of which included very specific ADCs with an identified loved one who is deceased. In fact, most of the participants cried during the experience and reported "feeling better" afterward. At the time of this writing, the data is still being analyzed but looks promising as a tool for both personal exploration and grief therapy.

Obviously, a VR experience has several distinct advantages over a traditional psychomanteum. First, it increases accessibility, making the only requirements access to a VR headset and the specific program. In addition, the VR version of the experience is only twenty-five minutes long , whereas a traditional experience is forty-five minutes at a bare minimum. For more information about obtaining or working with a VR psychomanteum experience, send an e-mail to info@neuromeditationinstitute.com with the subject line: VR Psychomanteum

CHAPTER EIGHT

PSYCHICS AND THE PSYCHEDELIC BRAIN

SHORTLY AFTER MOVING into our current office, I noticed a new coffee shop a few blocks away. What caught my eye were the signs and flags hanging in the windows. There was an LGBT/Trans Pride flag, flags from various countries, and signs written in sign language (ASL: American Sign Language). This created a very inclusive and accepting vibe that I was instantly drawn to: everyone was welcome. When I had a chance to go there for coffee, I immediately noticed the merchandise counters near the front door. There were crystals and gems, tarot decks, and various books about spirituality and occult topics. What had I stumbled in to? I struck up a conversation with the woman behind the counter and learned that she was also the owner, having recently moved to Eugene, Oregon, from Portland. Turns out Julie Rost (whom you met briefly in Chapter 6) was a psychic medium, tarot reader, and energy healer. She also taught psychic development classes at the coffee shop. Because I tend to have a skeptical nature and don't necessarily trust that everyone who claims to be psychic or a medium is truly gifted in these abilities, I wanted to check her out. After a few coffee shop conversations, I invited Julie over to the office for a psychic-brain mapping "play date."

Before diving in, we spent a little time getting acquainted. Julie told me that her mother and grandmother both had psychic abilities but came from a strict religious background and were not allowed to explore that part of themselves. She indicated that, for most of her life, she did the same thing—denying or avoiding her abilities. It is when Julie stopped using cannabis and alcohol and developed a disciplined meditation practice that her gifts strengthened, and she stepped into them fully. That was fourteen years ago.

When I asked Julie how the psychic process works for her, she described it as dialing a radio receiver until she finds the frequency of the other person or spirit. Once she finds it, she can go back to it rather easily. She suggested that this may be the same process that happens for spirits. They look for a signal match and then "come through." Since part of my journey is to develop my own skills, I asked if she had any advice for beginners. Julie said the most important thing is to listen and trust your intuition. Following that, she indicated that it is important to learn to quiet your mind and stay relaxed.

Tuning in with Crystals

Because this was our first official meeting, I left the format open and flexible. I told Julie she could do whatever she wanted, and we would just collect some EEG data to see what was happening in her brain. She indicated that she often did her work while holding various crystals and stones. Crystals and specific gems are believed to hold vibrational qualities that can help with healing, concentration, and creativity. Because they are composed of stable patterns of molecules, they are thought to help with channeling or directing energy. Julie brought a few of her favorites with her, curious to see if they showed different patterns in her brain. The spontaneous way this testing emerged was almost like speed dating. After I recorded a baseline EEG, Julie would hold one of her crystals and tune in, verbally reporting whatever was happening. After about one minute, we would pause, and she would switch crystals and repeat the process. This by itself was impressive. It suggested that she did not require any preparation time, meditation, or prayers; she could just drop right in.

She began by holding a multifaceted, pure quartz egg. She described seeing a woman standing behind me and a small dog going around her feet. This person was wearing a medicine pouch that had something in it. This entity communicated to Julie (and Julie communicated to me) that I was "on the right track." I assumed this meant the work I was doing attempting to understand and explore the psychic brain. The being encouraged me to "look at the big picture" and "avoid rabbit holes." This was sound advice as I tend to get excited about a lot of things that can draw my focus away. Julie indicated that this experience felt more like mediumship.

Julie switched to a blue quartz stone. She said, "There's not much going on here." She told me I should "dot my I's and cross my T's." I'm not sure what that was in reference to, but it seemed to be another suggestion to stay focused and clear in what I was doing. Julie also mentioned that she was connecting with the person who gave her this particular stone.

The third stone was a Lemurian quartz crystal. (Lemuria was a legendary lost continent believed by some to have sunk to the bottom of the Indian Ocean). As soon as she picked it up, Julie immediately

said, "You are on to something," and then said, "If you want to remember something, use crystals. They can help with information."
She went on to state that it is possible to use crystals in creative and
innovative ways.

The purpose of this experiment was to see if Julie's brain waves
changed in relation to which crystal she was holding. This idea was
especially interesting given that she only spent about one minute
with each crystal. Based on a typical brain response to task switching,
you would not expect that this would be enough time to shift brain
wave activity in any significant way, much less showing a difference
between one-minute intervals.

EEG Analysis

To begin, I collected some baseline EEG data while Julie was sitting doing nothing with eyes closed, then with eyes open, and then
while having a relatively mundane conversation. Then I recorded her
while she held each crystal. Later, I analyzed the data by comparing
the baseline EEG data with the crystal data. This produced a series
of brain maps that show the difference between the two recordings.
Areas that are darker indicate a significant increase of activity in that
particular EEG band, while areas that are lighter indicate significant
decreases of activity.

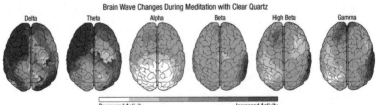

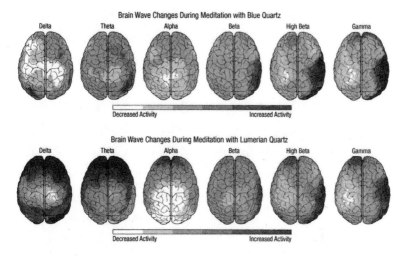

When you look at this head-to-head comparison, the results are pretty interesting. The first thing I noticed was that the pattern for the fast brain waves (beta, high beta, gamma) were similar across all of the crystal conditions. There was increased activity in the prefrontal cortex, as well as on the right temporal/parietal area. This is the same area implicated in Janet's shamanic channeling (see Chapter 1), as well as during spirit communication (see Chapters 2, 3, and 7). At the same time, there was a decrease of activation in the left central/temporal region. This shift of activation to the right side suggests more of a nonverbal, emotional, embodied way of perceiving information.

What I found even more interesting was the difference in the slow brain waves. There was a significant increase of delta and theta—the slowest brain waves—with the first and third crystals, while the second crystal showed a decrease of delta and no real change in theta. You might recall that Julie described that there "wasn't much going on" while holding the second crystal, the blue quartz one. This suggests that the big changes in delta and theta may have been associated with a perceived connection or feeling more tapped in. There were also some differences between the crystal conditions. The increase of slow brain waves during the pure quartz recording occurred almost exclusively in the left hemisphere of the prefrontal cortex, while the Lemurian crystal showed an increase of these brain waves in both hemispheres. It is not clear what this might mean. However, Julie reported that the first recording (pure quartz) felt more like mediumship, whereas the third recording (Lemurian quartz) felt more psychic. This may provide additional evidence that these two states (mediumship and psychic) differ in specific ways that require somewhat different brain activation/deactivation patterns. Perhaps different types of crystals are better aligned with different types of psychic activity. I also observed that this was not the first time we have seen an increase of slow brainwaves during certain psi states, particularly in the theta frequencies.

The Role of Theta

Because theta dominance in adults can create a sort of dreamy, trance-like feeling, similar to the sensation just before falling asleep, it can sometimes be difficult to recall the insights or messages received when in a purely theta state. For this reason, it is often necessary to simultaneously develop some faster brain wave activity. Because we have all of the brain waves, in all brain locations, at all times it is possible to increase more than one brain wave at the same time and/or to "adjust" the behavior of our various brain waves as a way to transfer information from the subconscious (theta) into the conscious (beta) mind. Perhaps this is what many of the psychics and mediums are doing? While they are increasing theta, they are often simultaneously increasing faster brainwaves such as gamma. This may be the brain's strategy to tap into that broader base of shared information (theta) and retain it in the conscious mind (gamma).

The final part of the crystal EEG analysis that I found intriguing was the dramatic decrease of alpha brain waves seen in recordings one and three. This decrease was seen in the center of the back of the head, the hub of the default mode network (DMN). You might recall that we saw this pattern show up in my brain during the psychomanteum experiment (see Chapter 7). Having worked in the field of psychedelic medicines, I was aware that this pattern—a decrease of alpha in the DMN—was the most consistent brain change seen during a psychedelic state. Surely this wasn't a coincidence. Given that psychic development and psychedelic medicines are both two of my passions, I wanted to dig a bit deeper to understand this connection.

The Brain on Psychedelics

The region of the brain most impacted by psychedelics is the default mode network. As the name implies, this is not so much a region as a set of regions that work together to accomplish certain tasks. The primary function of the DMN seems to be related to the creation of a sense of self, ego, or identity. In fact, when the DMN is doing its job, we are typically thinking about ourselves or thinking about ourselves in relation to the world. Every time you are worrying about the future, remembering the past, making a mental list, thinking about your family, deciding what to eat, planning a trip, or any number of other cognitive processes that involve you as a character in the narrative, you are engaging the DMN.

In a psychedelic state, this network gets quiet, blood flow decreases, and brain waves settle down, essentially shutting down normal functionality, and allowing the opportunity to see and experience ourselves in a new and different way. In addition, communication between the areas that make up the DMN also relaxes. This decrease in functional connectivity within the DMN has been referred to as "network-level disintegration" (Speth et al., 2016) and is associated with the experience of ego dissolution (Nour et al., 2016). When you turn down or disrupt the normal functioning of the DMN, this leads to a significant change in how you see yourself. If this process is taken to an extreme, as can occur with high-dose psychedelic sessions, it can lead to feelings of "oceanic boundlessness" of unity with the Universe (Roseman et al., 2018).

Interestingly, while connectivity within the DMN decreases, connectivity between the DMN and other brain networks increases during a psychedelic journey. This expanded communication between regions appears to be related to increased psychological flexibility and brain complexity (Carhart-Harris and Friston, 2019). Essentially, this process enhances the brain's ability to change and adapt, to take in new information, and to consider new possibilities. In essence, you are increasing the entropy or chaos in the brain, which makes it more flexible.

Entropic Theory of Consciousness

Carhart-Harris, a leading researcher in the field of psychedelics, has articulated the theory that consciousness exists on a continuum of entropic states* (2014). This model posits that states of consciousness represented by high levels of entropy become increasingly flexible and random. At the extreme, this can represent high levels of disorder, such as might be seen in early psychosis, dreaming, infant consciousness, or psychedelic states. Low entropy implies highly

*The term entropy originally comes from the field of thermodynamics to describe the amount of energy available to do work.

ordered but inflexible cognition, as might be seen in a variety of mental health concerns including depression, OCD, and addictions (Carhart-Harris et al., 2014). According to this theory, psychedelics provide therapeutic value by introducing entropy into a rigid system, creating a relaxation of previously held beliefs (Carhart-Harris and Friston, 2019) and a potential opportunity for healing. They argue, quite persuasively, that the DMN is generally organized to limit information in an effort to maintain a consistent and coherent sense of self. In this way, the normal consciousness experienced during regular waking reality can be considered somewhat rigid.

Based on this notion, increasing flexibility in the brain could potentially facilitate the development or uncovering of psi abilities. We saw an example of this in Chapter 1 in which Janet developed the ability to channel shamans in response to an altered state experience. It also seems that some psychics and mediums, like Julie, may be able to naturally shift into a state of increased entropy as a way to access information beyond the self.

I was curious if people who reported psychic and mediumship experiences demonstrated more neural flexibility than those that did not report these abilities. It made sense and almost seemed like it should be a prerequisite. If your nervous system is stuck in a rigid way of perceiving the world, it seems unlikely that you would be able to access information and communications that exist on a more subtle level. To test this hypothesis, we conducted a study to examine this relationship.

Psychological Flexibility and Psychic Experience

To examine the idea that psychological flexibility is related to psi abilities, we surveyed 302 research participants recruited through social media and the Psychic Mind Science website (www. psychicmindscience.com). The survey was comprised of three different self-report scales, including the Australian Sheep-Goat Scale (ASGS, this is a commonly used questionnaire to examine beliefs in the paranormal.) The name comes from the idea that believers are "sheep," while non-believers are "goats."), the Multidimensional Psychological Flexibility Inventory (MPFI), and the Highly Sensitive Person scale (HSP). The ASGS scale asks respondents about their belief and experience with a variety of psi-related activities that are focused on ESP, life after death (LAD), and psychokinesis (PK, the ability to move things with one's mind). The MPFI is, as stated, a measure of psychological flexibility, and the HSP is a measure of sensitivity to both internal and external sensory stimuli. We examined the relationship of these scales to identify any significant correlations. My hypothesis was that the experience of psi would be more strongly correlated with psychological flexibility than simply a belief in psi. My logic was that belief requires less flexibility than experience. Some might argue that belief could be suggestive of more rigid states of consciousness (see Robert Anton Wilson's book, *Cosmic Trigger, Vol. 1)*. When we examined the data (see next figure), we discovered highly significant relationships between *all* of the scales. In fact, there was less than a 1 percent chance that the relationships were due to chance. Not surprisingly, the strongest correlation was between belief in psi and psi experiences (r = .816). Note: correlations are indicated by a number showing the degree of similarity. The closer the number is to 1.0 the higher the similarity. The next strongest correlation, as predicted, was between psi experience and psychological flexibility (r = .340), followed by belief in psi and psychological flexibility (r = .227), then high sensitivity, which was nearly identical in relation to psi beliefs and experiences (r = .194 and r = .198, respectively).

Correlations Between Psi Beliefs, Psi Experiences,
Psychological Flexibility, and Sensitivity

ASGS (Australian Sheep Goat Scale)	BELIEF (Belief in Psi)	EXPERIENCE (Experience of Psi)	TOTAL
MPFI (Multidimensional Psychological Flexibility Inventory)	0.227	0.340	0.295
HSP (Highly Sensitive Person Inventory)	0.194	0.198	0.206

All significance values are p <.001

Based on these results (and common sense), it seems that those reporting an experience of psi generally have more flexible/higher entropy brains than those with no psi experience. It also seems clear that psychedelics create more entropy. However, there was still not a clear connection between psychedelics and psi experience. Was there any more direct evidence that using psychedelics could enhance or influence psychic abilities? As I considered this question, I immediately thought of my friend Kat Keiser. While I didn't know Kat well, I knew that she has been involved in psychedelic medicine work for many years, and I also knew that she practiced nature-based *magick*. I wanted to hear her thoughts on the subject. When I first reached out to her about these questions, I sent a message on Facebook. Here is part of our exchange:

Hey there, just checking back in. Also, question. Do you think that working with psychedelics has increased your psychic-type abilities (intuition, spirit communications, etc.)?

The answer to your question is an undeniable YES. I absolutely believe my work with psychedelic medicines have opened my psychic channels to intuition and spirit. In fact,

Okay, seemed like a pretty direct, affirmative (and emphatic) response. Kat was gracious enough to come in for an interview to discuss this a bit further.

Kat

I think of Kat as a modern-day spiritual monk. She has dedicated herself to her spiritual development and created a lifestyle to support this. In addition to daily yoga and meditation, she does not use alcohol or caffeine and is very careful with her diet, eating mostly raw foods. She fasts every month for three days and sometimes for more extended time periods. She spends a tremendous amount of time in nature. She trains with and works with indigineous, shamanic tribes, and incorporates the regular use of plant medicines into her personal practice. In addition to intensive peyote ceremonies, Kat indicated that she microdoses with psilocybin and/or peyote several times each week. She indicated that these medicines help her maintain an open heart and a connection to Spirit. She told me that she considers these medicines as training wheels that are given to us by the Earth to help us open up portals to our Higher Selves.

She gave me numerous examples of times when she experienced telepathic communication while using these medicines. Most recently, she was in Mexico and was scheduled to attend a nine-day peyote ceremony. She was going to cancel as she was experiencing significant neck pain and the ceremony was physically demanding, including extended hikes through the desert carrying all of her camping gear. Someone in the group had training as a chiropractor, and she considered asking him for an adjustment before the journey, but decided against it. If something went wrong during the manipulation she had no access to a hospital or medical care. Despite the discomfort, she decided to participate in the ceremony. She had been preparing for quite some time and felt the pain would be worth it. On the second day, Kat was still not feeling right. She had a lot of internal tension and felt agitated. The group was sitting for a meditation and Kat had a clear impression that she should approach the chiropractor and ask for help. Within moments of having had this thought, the chiropractor walked over to her and spontaneously said, "Now is the time." How does this kind of nonverbal communication happen? Somehow he had picked up on her thoughts and felt compelled to help. He used a combination of chiropractic techniques and energy healing which led to Kat "seeing" all of her tension and stagnant energy leave her body. Afterwards she felt significant relief—lighter, more energized, and pain-free. This may have been a coincidence. Yet, when you consider the circumstances and take it in context of other stories, it supports the idea that somehow Kat and the chiropractor were telepathically in tune with each other. They seemed to be able to commuicate energetically, without words, facilitated by the peyote.

Kat gave many other examples of telepathic experiences she has experienced with the plant and animal worlds. She indicated that there have been dozens of times where she would intuitively know exactly what a plant needed to regain its health, or what kind of medical problems a particular animal was experiencing. In all of these cases, when she listened to this inner guidance, she was shown to be correct. Kat also described times that she had premonitions of things that later happened, including detailed visions of car accidents.

When these experiences first began happening, Kat indicated that she didn't trust them. She assumed she was just making things up or seeing connections that were not really there. However, after dedicating herself to deepening her relationship with her spirituality, she has come to fully trust these guidances when they come through.

According to Kat (and many ancient traditions), everything is connected on an energetic, spiritual level. Yet, we are here, in this human existence in a very dense form, believing that we are isolated and alone. When we are able to transcend this false sense of separateness, then we have access to much more information. Everything, all past and future information, is available if we can only get beyond our self-limiting beliefs and behaviors to tune in. In order to remember and connect to the Oneness—the single consciousness—Kat indicated that it is necessary to maintain a certain level of "activation" and receptivity. She said that in order to do this it is important to shift our energy and attention away from our baser instincts and behaviors. If we spend all of our energy thinking about money, food, and sex there is little energy left to activate higher states.

While Kat is using more esoteric language and ideas, the shift in consciousness she is describing sounds very much like a shut down of the right parietal lobe (RPL) and the DMN. Remember, shutting down these areas, or shifting their function significantly, appears to change people's perceptions of self in relation to others. It breaks down boundaries and increases feelings of empathy, connection to others, and spirituality. It allows an opportunity to experience a unity consciousness. Of course, if all information is always available, then our task is to get out of our own way. According to this understanding, we are essentially blocking ourselves from this information because of our self-limiting ideas of who and what we are. Psychedelic-type experiences can potentially provide access to this Oneness, helping us remember who we are outside of our limited identity.

Jeff and Ketamine

Overlapping with my work on this book, I was involved in a year-long training program to facilitate psychedelic-assisted therapies. Even though the State of Oregon (where I live) had voted to legalize the use of psilocybin when working with a trained (and state-licensed) facilitator, this program had not yet been implemented at the time of this writing. Instead, I was working with a couple of physicians to offer ketamine-assisted therapy (KAT).

Ketamine was approved by the FDA in 1970 for use as an anesthetic. In fact, it is identified as an essential medication by the World Health Organization due to its usefulness and strong safety profile. When used in lower, subanesthetic doses, it can create a significant level of dissociation, allowing the person a form of separation from their ordinary reality and usual sense of self. Used in an appropriate context (set and setting), this medicine has been found to be effective in the treatment of treatment-resistant depression, pain, addictions, obsessive-compulsive disorder, and various other mental health concerns. Some of these effects appear due to the medicine's ability to increase entropy and psychological flexibility. As part of my training, as well as for personal growth and exploration, I have personally worked with this medicine several times. One of these journeys was shortly after my conversation with Kat, and I was struck by the similarity of her perspective and my experience.

The Journey

After creating a pleasant and comfortable space in my spiritual practice room, I placed the chopped up ketamine troche in my mouth and maneuvered the pieces under my tongue. A *troche* is a kind of waxy tablet that dissolves in the mouth. This is used for sublingual administration. Essentially, the ketamine enters the

bloodstream through the blood vessels in the mouth, which are particularly receptive under the tongue. For the most effective session,
it is recommended to hold the saliva in your mouth for fifteen minutes or so to allow the troche to fully dissolve and for the maximum
amount of the medicine to enter your system.

After putting the ketamine into my mouth, I put on my headphones, turned on a Spotify playlist I created especially for ketamine
journeys, put on my Mindfold eyeshades, and laid back. The medicine is a little numbing in the mouth, and it can be uncomfortable to
hold a mouthful of saliva for an extended period, but I was already
beginning to feel a bit altered even before the fifteen-minute time
limit expired. I spit out the saliva and laid back, relaxing into my
"Moon Pod" beanbag chair. It did not take long before I was deep
into a psychedelic journey. I remember being a bit surprised as I had
taken this amount of ketamine before with only a mild psychedelic
experience. For whatever reason, this experience was much more
intense.

I very quickly lost track of my body. Actually, this is not even
a fair way to describe the experience. It is more like I forgot that
I had a body. My consciousness was moving through a variety of
flowing experiences, all directed by the music. In this state, the
music and the visual experiences felt intimately connected. It was
as if the music directed the movement and flow of energy I participated in. The experience was generally pleasant or neutral as my
consciousness observed the interconnectedness of various abstract
energies, flowing into each other, influencing each other, balancing
each other in a dynamic, organic way that felt very much that I was
observing the "real" nature of reality. Everything was ultimately this
dynamic flowing of energy. It was all vibration at different frequencies, morphing into various landscapes and figures before melting

again. Every moment that appeared in this timeless space existed and then was reabsorbed, shifting into something else. In these moments, I understood that everything is part of this process, including each of us. The idea that we are all separate identities, the importance we place on our selves, our jobs, and our status are all ridiculous in this context. In reality, this physical form will die, and our consciousness will return and join with this primordial energy pool before eventually becoming part of something else. As I began returning to my body, I was able to shift back and forth between this more fluid, basic reality and my manifested form as "Jeff." I realized that what Kat described was true. Everything is connected and the key to psychic, PK, energy healing, or spirit communication skills is to understand this reality and learn to move beyond the limited consciousness that is temporarily manifested in our human form. From this space, I didn't want to come back. I didn't want to be limited by my body and my thoughts. I knew that I would be sucked back into a dense version of myself in which my realizations felt distant and somewhat unreal.

If that all sounds kind of intense, it should, because it was. It was not scary. It was actually comforting. It was like seeing behind the curtain in the Wizard of Oz, getting a glimpse and an experience of how things really work, and understanding that our consciousness is much larger than we can imagine. In fact, it is common for people having an intense psychedelic experience to report that the experience felt "more real than reality." I would take it one step farther. I believe that experiences such as the one I just described are touchpoints into a higher reality. These realities are real. It is not just something the mind makes up, it is the "Truth." You know when you understand something deep in your spirit—down in your bones? This is that feeling, and it is no wonder these experiences can be life-altering.

While I did not specifically have a telepathic experience or some type of spirit communication, I did get to see beyond the veil. I was able to see how these abilities might work—why it is possible to communicate with noncorporeal beings or travel through time, or to know something you shouldn't be able to know. This experience and understanding demonstrated to me the importance of openness, flexibility, trust, and letting go. These psychological/mental capacities may be a prerequisite to accessing our psi potential.

Next, I wanted to know if these ideas had been studied before, to know if there was any scientific evidence that psychedelics could influence psychic abilities.

Altered States and Psi

The connection between anomalous experiences and altered states of consciousness is well-established, particularly in relation to the use of psychedelics. With minimal digging, I found several studies that seemed to support this connection. Here are the highlights.

- Sixty-two percent of those using *ayahuasca* (pronounced eye-ah-wah'-ska, a South American, plant-based psychedelic brewed as a drink) reported experiencing an OBE during one of their psychedelic journeys (Luke, 2009).

- Those who frequently use psychedelics are significantly more likely to report having had an OBE than nonpsychedelic users (Luke, 2019).

- Near-death experiences are relatively common during certain psychedelic experiences, particularly with the use of 5-MeO-DMT (a psychedelic found in certain plants and some toads) or ketamine (Luke and Kittenis, 2005; Corazza, 2010).

- Encounters with some form of discarnate entity are estimated to occur in about half of all high-dose DMT experiences (Strassman et al., 2008).

- Experiences of telepathy during the use of cannabis has occurred in up to 83 percent of respondents (Tart, 1993).

Of course, not everyone interested in gaining cognitive/psychological flexibility is interested in working with psychedelic medicines. Luckily, there are many other nonmedicine approaches to inducing a psychedelic-like altered state, including vibroacoustic therapy, float tanks, and certain styles of meditation (breathwork was discussed in Chapter 1).

Try It Yourself

Vibroacoustic Therapy (VAT)

This is a type of sound therapy that translates the frequencies of sound into vibrations that are then felt throughout the body. This process can be accomplished through a range of specially designed massage tables, mats, chairs, backpacks, or wristbands. While these devices are generally used to induce a state of relaxation, they are also used for a wide range of physical and mental health concerns. It is unclear exactly why this approach seems helpful for so many people, although it is suspected that the bodily vibration increases vagal tone, leading to a significant relaxation response. When the vibration is paired with certain kinds of music it can lead the user into a deep, meditative, trance-like, or even psychedelic state. This is a safe approach and is generally very pleasant. Some of the devices we recommend include the Sensate vibroacoustic pod (getsensate. com), inHarmony meditation cushion (iaminharmony.com), and the subpac vibroacoustic backpack (subpac.com).

Sensory Deprivation Tanks

Sensory deprivation tanks entered the public consciousness in 1980 with the film *Altered States*. The film follows a scientist who begins experimenting with other states of consciousness, believing them to be as real as waking consciousness. The primary method of investigation involves spending time in a sensory deprivation tank, often aided by a variety of psychedelic drugs. As a sci-fi/horror film, the story gets a bit weird as the main character enters deeper and stranger states of consciousness leading to his actual physical transformation into more de-evolved species.

The film was partially based on the work of Dr. John Lilly. Lilly was a neuroscientist, psychoanalyst, philosopher, and *psychonaut*. He is credited with creating the first sensory deprivation tank in 1954 as a means to reduce as much external stimulation as possible, allowing an unfiltered experience of human consciousness (Lilly, J. C., 2007).

The sensory deprivation tank experience involves stripping down and getting into a self-enclosed pod that is partially filled with skin-neutral (93.5 degree) salt water. There is so much salt in the water that you float effortlessly, as in the Dead Sea. When the pod closes, all sound and light is blocked out. In this environment you have access to very little of your typical sensory experience. There is nothing to see, hear, or feel. In fact, your body doesn't even have any weight. A typical float session lasts between forty-five to ninety minutes and can lead to deep feelings of relaxation. For this reason, they are often used therapeutically for the reduction of pain and/or stress. For some people, this float experience can also lead to altered states of consciousness including recalling old memories, internal "journeys," profound insights, intense feelings, and, in some cases, hallucinations.

As part of a qualitative study conducted by Kjellgren and colleagues (2008), eight floaters were interviewed about their experiences. Many of them described a relaxed attention that allowed their mind to simply wander, but several of them also described significant anomalous experiences, including hearing internal voices or music. The more profound experiences reported included participants describing the feeling of being transformed into a bird and flying. Another participant described a perinatal experience, as if they were in a fetal state in a uterus and unborn. Other participants described having out-of-body experiences and feeling transported to another world, as if they were floating in space.

Curious about trying it? Most mid- to large-sized cities in the United States have float centers or spas with a float tank that is available. Simply googling "float tanks near me" should provide you with some options.

Deep States Neuromeditation

A recent survey of 3,684 people found that consistent meditators were significantly more likely to have had a paranormal experience compared to nonmeditators (PRWEB, 2022). For example, of those surveyed, 34 percent of the meditators reported having seen or spoken to an angel or spirit guide, while only 10 percent of the nonmeditation group endorsed the same question. Similarly, 44 percent of the meditators had a premonition of something that later occurred, while only 20 percent of nonmeditators reported such an experience. Other research on meditation has shown that certain styles of meditation (such as Vipassana) increase entropy within the alpha and gamma brain wave bands (Vivot et al., 2020). Meditation and psychedelics also share other brain changes, including decreased activity in the default mode network (Carhart-Harris et al., 2012; Berkovich-Ohana et al., 2016).

Styles of meditation differ based on how you are directing your attention, your intention, which brain waves are involved, and in which brain regions they occur. Using brain wave biofeedback along with meditation, we have created an approach that guides the brain into a state that mimics what is seen in research with psychedelics, quieting the DMN. The system provides auditory feedback through the volume of the meditation music to let you know if you are on the right track. This feedback can be used to help you learn how to drop into a deeper state of consciousness or an altered state. After one experience at our clinic the client said, "I felt like I was flying very fast. I would slow down, viewing the Milky Way from different viewpoints, seeing the planet throughout humanity's progression. I felt like I was remote viewing, experiencing parts of something, yet parts of nothing. All of this was happening at once and yet not at once." This person's experience indicates that it is possible to enter an altered state through meditation; in this case, EEG-guided deep states meditation. If you are curious, visit www.neuromeditationinstitute.com to explore the most recent advances in this technology.

CHAPTER NINE

PSYCHOKINESIS

ON NOVEMBER 27, 2013, Janet Mayer sent me an e-mail message with the subject line "Telekinesis." The opening sentence read, "So, my question is: Who is going to be able to do this first? You or me?" The message included a link to a YouTube video of a guy moving a pencil with no signs of physical contact. He would meditate for a few moments with his eyes closed, open his eyes, and hold one hand up toward the pencil, which was lying on the table. He would gently move his hand out toward the pencil, causing it to roll forward a bit. He could also reverse course and draw the pencil toward him by pulling his hand back. It looked as if there was an invisible tether between his hand and the pencil. I loved it.

I had been interested in the idea of telekinesis since childhood but had never seen it outside of a movie. Of course, I realize there are a thousand ways to manipulate this situation. Aside from camera or editing trickery, it is possible that the person in the video was able to tilt the table, or maybe his hand movements were creating enough air pressure to exert the effect. Stage magicians do far more impressive tricks like this all the time. In fact, a good performer can be standing right in front of you and there is no way for the untrained eye to figure out how they make something disappear or levitate. Despite this awareness, the video was impressive and believable, not just the appearance, but the guy's attitude and approach. He did not seem overconfident. He wasn't making outrageous claims. He wasn't doing anything dramatic like closing doors or moving furniture around the room; he was simply rolling a pencil a few inches. He wasn't selling anything. He also clarified that the video had been edited to only show the successful trials, acknowledging that it didn't always work and took a lot of patience. He even discussed the different ways that people may think he is cheating and safeguards he had in place to prevent other influences from impacting his work.

Other YouTube videos showed people moving a sort of pinwheel made of aluminum foil. This made more sense to me. If you are new at trying to move objects with your mind, that is, psychokinesis or PK, you probably want to start with something that is very light and easy to move. While I suspect that the ability to move objects with your mind has very little to do with the physical manifestations of the object, it does seem important to believe you can move the object. There is a famous scene in Star Wars Episode V: *The Empire Strikes Back*, where Luke Skywalker is on the Dagobah planet with Jedi Master Yoda learning to harness "the force." Luke is easily discouraged (and a bit whiny) and tells Yoda that he cannot move the X-wing Starfighter, which is buried in the swamp because it is "too big." Yoda replies in characteristic fashion, "Size matters not," and goes on to give a mini lecture about the importance of recognizing the way the force connects everything. Yoda's description of the force sounds a lot like descriptions of qi or prana: a life force that moves within and between each of us. If you can learn to harness this energy and move past the limiting beliefs of the mind, then it should be possible to move any object.

First Attempts

Because I recognized that my mind would reject anything that seemed difficult from a physical standpoint, I decided to work with a ping-pong ball because they are lightweight, they roll, and have minimal resistance to a flat surface. I sat at my kitchen table for fifteen or twenty minutes, several times a week, staring at the ball, directing all of the focused attention I could muster toward the ball with the intention for it to roll forward. When this didn't work, I would slowly move my hands toward the ball or shift into an external qi healing attitude, holding space around the ball and "asking" it to move forward. None of this worked, which was extremely discouraging. How many hours can you sit staring at a ball with nothing happening before you become convinced that it is not possible? In my case, about two weeks. Despite my frustration, this experience did lead to an important insight regarding my telekinesis attitude. I was trying to use energy or qi to exert a force on the ball. This way of doing things, while potentially possible, relies on a mechanistic approach. I might as well have tried moving a chair by blowing on it. I was trying to use an invisible force (my mind) to impact a physical object, which, in hindsight, seems rather silly. This wasn't about physical force or a Newtonian view of the Universe. If it truly is possible for humans to influence physical objects with their minds, then I needed to change the way I was considering the task.

The Role of Belief

This realization took me back to that initial e-mail from Janet. In addition to her telekinesis challenge, she had included a blog post written by Andrei Bannikov, Founder of Binaural FX, titled, "Teleke-nesis—a Theory That Makes It Possible."* In this article, within the first paragraph, he states, "To master an ability to influence kinetic properties of an object, it is imperative to change your perception and understanding of what that object is." He goes on to describe the necessity of viewing the object as something more basic than a phys-ical object. Is it possible to shift your perception of the object in such a way that you can tune in to the *energy* of the object rather than its dense *physicality*? If you are trying to move something physical with mental thoughts, that is impossible. However, if you are attempting to shift energy with energy, that sounds more feasible (and makes more sense). This idea reminded me of a scene from the first Matrix movie. Neo, played by Keanu Reeves, is waiting to visit the Oracle. In the waiting room he sees a young boy with shaven head, wearing a monk's robe. The young boy is holding a spoon and causing it to bend simply by looking at it. He hands the spoon to Neo and says, "Do not try and bend the spoon, that's impossible. Instead, only try to realize the truth: there is no spoon. Then you'll see that it is not the spoon that bends, it is only yourself." Similarly, moving objects with your mind requires you to reframe your perceptions. Do not try to move the object with your mind; instead, connect your mental energy to the energy of the object. If you can accomplish this shift in perception, you only need to move the mind and the object will follow. This all sounds great and semi-logical, but how do you con-vince a brain that has decades of rigid training to perceive and con-nect to invisible energy fields?

An updated version of this blog can be found here: https://conscious-reminder.com/2016/11/26/telekinesis-theory-makes-possible/.

Practice, Practice, Practice

Bannikov describes this process as "sensing into the energy field of the object you are trying to move," that is, tuning in to its vibration, dynamics, or aura. If you simply stare at the object, not much is likely to happen. If, however, you can sense and connect to the energetic nature of the object, then it will respond to your intentions. If you already have experience with energy fields and auras, this shifting of attention may not be too tricky. On the other hand, if this is a new idea for you, you may need a bit more guidance.

Bannikov talked about building this "sixth sense" by beginning with the physical movement. Literally, he described physically holding the ping-pong ball and pushing it forward and backward on the table until he was able to internalize the feeling of the ball moving on the table. This is a clever approach to helping move from the physical to something more subtle. With enough repetitions, it increases the possibility of recreating the feeling in your mind without touching the ball or moving your body, so that you use only your sensory memory.

I practiced this way for a while, but without success. However, this practice led to several more insights into this work. First, I recognized that some part of my mind or personality did not truly believe that I could do this. I *wanted* to do it. I *believed* it was possible, but somehow, I did not fully believe it was possible *for me*. Maybe this limiting belief (however subtle) was blocking the process. I also understood that a significant amount of practice may be necessary to develop my telekinesis abilities. And why shouldn't it? Learning to play an instrument, speak a second language, or meditate all take practice and repetition. Why would this be any different?

To address these insights, I first switched to an easier target for my experiments. Rather than attempting to roll a ping-pong ball on a table, I switched to a pinwheel made of aluminum foil. Essentially, this is a square of aluminum foil, folded twice. When it is unfolded, there are four creases coming from the center point, making a tiny umbrella shape. This pinwheel is then balanced on some kind of stand—I put a nail through a rectangular-shaped pink rubber eraser. While it may not make any difference in the energy world, moving a thin piece of aluminum foil seemed far easier than moving a ping-pong ball. I was basically switching to a target that was easier for my mind to accept. I also decided that I was going to keep experimenting with this until I got something to happen. Interestingly, I didn't have to wait long.

Shortly after switching to the pinwheel format, I noticed movement. Sometimes the tinfoil would just wobble a bit, sometimes it would turn slightly in one direction or another, and occasionally, it would make one or more full rotations. This was exciting, yet I still didn't fully trust it. The movement was inconsistent, and the tinfoil could easily be influenced by minor air currents in the room or heat from my hands. In fact, this awareness led to some helpful experiments. What would happen if someone walked by my table? What if the air conditioning kicked on? What if my face or hands were too close? These kinds of observations helped me to discern what real movement felt like and what artificial or artificially influenced movement felt like. After several months of this practice, I became pretty good at knowing which process was engaged simply by how the tinfoil was reacting. On the one hand, this increased my confidence. On the other hand, there were still too many uncontrolled variables that I could not rule out. I needed a test that would not be influenced by air. There was a solution that I had seen in a few

YouTube videos. It simply involved placing a transparent covering over the pinwheel. In my case, I used a large clear, plastic mixing bowl. Interestingly, as soon as I added this barrier, it felt like I had returned to the beginning. Nothing moved. No wiggles, no wobbles, and certainly no full turns. My first reaction was to assume that I had been kidding myself with my prior successes. That everything I had previously witnessed was simply an unknown air current doing the work for me. Despite these misgivings, I kept at it, and I noticed that the experience of working with the pinwheel felt different. I no longer felt connected to it when it was underneath the bowl. Somehow it felt far away. If I took the bowl off, I could usually get it to move. As soon as I replaced the cover, nothing. I thought maybe something about the covering was creating a literal barrier to the energy being used to move the pinwheel. Maybe glass would be different? I switched to a glass bowl and got the same result. Eventually, I realized that this was another layer to my limiting beliefs. Because there was now a barrier between me and the tinfoil, it became extremely challenging to believe that it was still possible. Eventually, with time, patience, and practice, I got the tinfoil to move under cover. Not often and not very much, but enough to show me that it was possible. I wanted to know more. Has anyone researched this? Are there people teaching this? What is possible besides moving aluminum foil and ping-pong balls?

The Research

As I started looking into research on psychokinesis (PK), I discovered that there are two different types of PK, often referred to as micro- and macro-PK. Micro-PK refers to the ability to influence very small things, usually smaller than can be observed with the naked eye. Macro-PK refers to larger influences, such as moving a

pinwheel or ping-pong ball. Most of the research in this area has focused on micro-PK impacts using random event generators (REGs).
You may recall from Chapter 5 that REGs are computer-based systems that can measure rapid oscillations between two equally possible binary states (e.g., either a one or a zero). Imagine flipping a
coin hundreds of times a second. There are only two options (heads
or tails) and, with enough repetitions, you should get heads 50
percent of the time and tails 50 percent of the time. Micro-PK researchers use such devices in their studies to test whether research
participants can influence these outcomes. Essentially, is it possible,
simply through thought and intention, to shift the occurrence away
from 50 percent for each outcome option, producing more zeroes
or ones?

While many researchers and labs have experimented with micro-PK using REG techniques, the most famous is the Princeton
Engineering Anomalies Research (PEAR) Laboratory, which was
active at Princeton University from 1979 to 2007 and focused on
parapsychology, specifically, psychokinesis and remote viewing. In
a typical PEAR-style experiment, the participant would sit across
from the REG and receive instructions to use their mental influence
to affect the outcome. The REG was designed to rapidly provide
random numbers between zero and 200. Statistically, 50 percent of
the values should be above 100 and 50 percent below 100; when
you average all of the values together, they should equal 100 (exactly in the middle). Participants would be asked to either increase
the number of high values (high condition), low values (low condition), or average value (100; average condition) using their intention. Testing for a single participant might include between 1,000
and 5,000 trials per condition being tested (high, low, average).
After twelve years of data collection with ninety-one participants,

the overall results indicated highly significant findings supporting the idea that participants could influence the outcome with their intention. For the high condition there was a 99.96 percent probability that the results were NOT due to chance. In the low condition there was a 97.63 percent probability that the results were NOT due to chance (Williams, 2021).

Beyond the PEAR studies, many other labs have explored micro-PK with varying rates of success. In an attempt to combine all of this data, researchers have completed a series of reviews designed to look at the cumulative data across all micro-PK research trials. In each of five separate meta-analyses, researchers have consistently concluded that the results against chance were highly statistically significant (see Radin and Nelson, 1989, 2003; Radin, 1985, 2009; Bosch et al., 2006). In essence, there is virtually no way that these results were random. This is all very cool and provides rather solid evidence that humans can influence nonhuman energies with their minds. Of course, it is also important to note that some participants (and trials) were much more successful than others. When you combine thousands of data points across multiple environments and participants, any of the extremes in the data get washed out and you lose some potentially valuable information. For example, what about the participants that did significantly better than others? What were they doing? What seemed to help?

In the book, *Real Magic*, Dean Radin discusses a series of micro-PK tests conducted in the 1970s and 1980s, which took the extra step of asking participants what kind of mental strategies they were using during their trials. In this way, they were able to match strategies with success. Radin reports that the most successful strategies included: resonance, asking entities for assistance, using emotion to help "power" the will, one-pointed concentration, physical relaxation, visual imagery, and talking to the RNG as though it were sentient. This information is gold. More important than the statistics proving that micro-PK is real, this helps us understand the mental attitude(s) that may facilitate this ability. Interestingly, but not surprisingly, some of this sounds a great deal like what Bannikov described. You might recall that he emphasized the importance of forming a connection to the object, which is essentially the same idea as resonance as described by Radin.

Trebor Seven

Sometime in my early explorations of PK, I stumbled upon a YouTube channel by a guy calling himself Trebor Seven. He describes himself as a self-taught practitioner and teacher of psychokinesis and has dozens of videos demonstrating his various PK abilities. Not only could he move a pinwheel (sometimes called a psi wheel), but he could also make it turn left or right and spin in full circles. He could do this when the psi wheel was under a glass case, and he was wearing a mask to prevent his breath from affecting anything. He had demonstrations knocking over a dollar bill that was folded and standing on its end. He could make it fall forward or backward. He could roll a Pringles can and cause ping-pong balls to move through a miniature obstacle course. After following several of his tutorials, I reached out to him through his YouTube channel and was surprised to receive a response. Turns out his real name is Robert (spelled backwards it's Trebor) Allen and he lived a few hours from me in Missouri. We talked a few times and discussed the possibility of mapping his brain while he performed some of his PK abilities. Robert was open to this experiment but warned that he wasn't always able to make things move and that the ability seems particularly fussy when he is being videotaped or asked to perform on demand. While skeptics may see this hesitance as evidence of fraud, I found his explanation to be sincere and consistent with my own experience. Robert explained that the mental state required for PK requires a significant level of relaxation and ego detachment. If there are any subtle stressors, anxieties, pressures, or similar influences, it is difficult to get the mind into a pure state of equanimity. Naturally, if someone is standing there waiting for you to do something that should be impossible, it will likely generate tension.

Energy, Mindset, and Empathy

Bannikov and Radin both indicated that resonance or connecting to the object was a key to PK success. Now, Robert was indicating that it was important to have a clear mental state as well. In an interview with Robert, I asked him if these two things were, in fact, the secret to PK. He answered with an emphatic "Yes!" and then added one more ingredient. After more than a decade of methodically developing his own abilities and teaching thousands of students, Robert indicated that his recipe for successful PK has three parts, which he calls energy, mindset, and empathy.

Energy

This part of the recipe is a bit difficult to articulate, but involves cultivating a refined, clarified, or heightened internal feeling of energy. Some might describe this as accessing a "higher vibration." This feeling often comes with a subjective sense of lightness or a "buzzy" sensation in the head or throughout the body, a sense of coldness (or warmth), an electric feeling, or tingling sensation. This type of activation is common to various forms of qigong and yoga, as well as to breathwork practices such as Wim Hof or SOMA. The idea is that you are using these practices to activate and clarify the life force, known by many names including qi, prana, and ki. (Note that some of this breathwork, such as Wim Hof, creates these sensations through hyperventilation.)

I found it interesting that energy cultivation was an important aspect of Robert's formula. This implies that PK might involve something more than just intention. It is not just a mental process; it may also involve energy. However, when I have asked if the PK practitioner is using energy to move the object, the answer is nearly always *no*. So, somehow energy is involved, but in an indirect manner. From my own experimentation, I suspect it has something to do with the connection between our energy field and our mental state. If you are in a low-energy state, such as feeling tired, distracted, or bored, or if your body is heavy and inflamed from lack of exercise, poor diet, or too much alcohol, this will affect your mental state and your ability to connect. In this way, raising your vibration appears to involve all aspects of lifestyle. A healthy, energized body supports a healthy, energized mind. Based on this understanding, it is not surprising that many high-level psi practitioners practice qigong or yoga, have a vegetarian or vegan diet, refrain from drinking alcohol, meditate regularly, and strive for psychological balance in their lives.

Mindset

After your energy is "ramped up," Robert indicated that the next step was to "slow your brain." At first, this may sound like a contradiction. How can you have an activated energy field and a slow brain at the same time? These two states actually complement each other easily and naturally. For example, one of the typical results of practicing qigong or yoga is a quieting down of mental chatter. For Robert, this mental state is about letting go of ego-based desires and shifting into a more receptive and open state of awareness. This doesn't mean that nothing is happening. Instead, it means that awareness has shifted away from thoughts about the self toward a more present-moment, gentle attention. Imagine a state in which the mind is alert, active, and aware, but relatively empty. You are entirely focused on the present moment without any thoughts of what will happen next or what happened before. This is the ideal state for PK.

From a neuromeditation perspective, this state sounds like a combination of three of the four meditation styles: focus, mindfulness, and quiet mind. The common denominator of these three styles is an inhibition of activity in the default-mode network. You might recall from previous chapters that the DMN is primarily focused on creating a sense of self. When this region is shut down it allows access to more subtle information, information beyond words, analysis, and judgment. This is the state we are seeking.

Empathy

The term empathy, in the way Robert uses it, seems to be the same thing as feeling connected or in resonance with the object. Robert described developing empathy as "getting really comfortable with the object." He indicated that he will often spend a great deal of time, using all of his senses, to fully explore an object, particularly if it is something he has not worked with before. What does it look like? Feel like? Even smell like? He indicated that he wants to become so familiar with the object that he can see and sense every aspect of it with his eyes closed. Robert also emphasized the importance of connecting to the *feeling of the action* you are asking the object to perform. Rather than visualizing the object spinning, connect with what it feels like to spin. If this is hard to recall, he suggests finding a merry-go-round and doing a few rotations. If you are attempting to knock something over by "pushing" or "pulling," remember that feeling. What is it like when you are leaning back in a chair and almost fall? Can you connect with that sensation? Essentially, you are attempting to find ways to bond with the target of your PK experiments.

Robert indicated that when all three of these ingredients are activated and in sync with each other, the PK work feels effortless. He described it as almost feeling "accidental." At one point he said, "We don't learn PK, we remember it," as if it was something that is natural to all of us—we have simply forgotten how to access it. If this empathic/quiet mind/high vibration state shows up in the brain the way I think, we should see a change of activity in the frontal lobes (focus), and a quieting of the DMN (focus, mindfulness, quiet mind). Unfortunately, Robert and I were never able to coordinate our schedules before I moved from Missouri to Oregon, so I never got an opportunity to see his brain in action. I did, however, find another PK expert willing to give it a try.

Caroline Cory

Toward the end of 2020, I noticed a movie on Amazon titled *Superhuman: The Invisible Made Visible.* It was an award-winning documentary studying people who can do "superhuman" things like move objects with their mind, change the pH of water, or see clearly while wearing a blindfold. I appreciated the approach because it appeared to be scientifically based. It tested these abilities with equipment that fit with my professional training. Initially, I was a bit skeptical as I tend not to trust things I see on TV. After watching the movie, however, I was convinced that the abilities demonstrated were authentic. These people appeared able to use their consciousness in ways that defied logic.

I looked up information about Caroline Cory, the interviewer and producer of the film. Turns out, she has several books and online training programs to teach some of these skills. I immediately signed up for one of her courses and reached out to her through her production company. I described my work and asked if she would have any interest in collaborating. A couple of weeks later I got a response. Initially, we explored changes in my own brain while doing a specific meditation she teaches called Connecting to Source. The results were intriguing and led to an in-person mini-retreat with Caroline and several of her advanced students. We spent a few days mapping everyone's brains while they were doing the Connecting to Source meditation, as well as during spirit communication. At the end of the weekend we mapped Caroline's brain while she attempted PK. We started the session with a paper pinwheel under glass. While I monitored the time stamp on the EEG recording, Caroline verbalized what was happening with the pinwheel. In this way, I could identify specific time periods when there was movement for later analysis. At around ninety seconds into the recording, Caroline said, "It's starting to react. It's wobbling." The wobbling continued off and on for another minute and Caroline indicated that she was having a hard time focusing. Her mind was busy from leading the weekend retreat and too much caffeine. We took a break and readjusted some of the environment, so it was less distracting. Caroline also did a short meditation to connect with her guides and settle her mind.

We started again—round two—and almost immediately Caroline indicated that the paper was wobbling again. This back-and-forth movement continued and shifted in intensity over the next few minutes. At various points the paper pinwheel would "jump" to the left or right, making a sudden partial rotation of anywhere from ten to thirty degrees. After about seven minutes, Caroline paused and removed the paper pinwheel, replacing it with a more traditional aluminum pinwheel. She indicated that it felt like the paper was "tired" and nonresponsive. She told me that she had been talking to the paper, attempting to give it energy, but wanted to try something different. The foil responded almost immediately, beginning with the back-and-forth wobbling pattern and then several small movements to the left. After a few minutes, Caroline indicated that she was finished.

As I contemplated what I had just witnessed, I realized that this simple demonstration was nothing short of a miracle. According to the laws of physics, this should be impossible. Something immaterial (consciousness) should not be able to influence something material (the pinwheel). These results were even more impressive when I remembered what Robert said regarding performing in public. He had warned me that this is tricky as the PK really requires a clear mind and an almost effortless engagement, which is difficult to achieve when someone is watching and waiting. With that in mind, it was impressive that we got any response at all. In fact, I wondered if my presence somehow also influenced the pinwheel. Even if Caroline was in the perfect mental state, was it possible that my anticipation, anxiety, and eagerness inhibited the process in some way? When I asked Caroline about this possibility, she explained that it is important to have a clear intention, blocking the energy of everyone else in the room. She told me that she always begins a session by visualizing herself and the PK target in an energy bubble designed to repel the signal of anything outside of herself. She said that if she hadn't done this ritual it is likely that I would have influenced her. As I considered all of this in relation to the brain mapping, I realized that I was really interested in the mental state associated with PK more than how dramatic the demonstration was, and we had plenty of time segments with active PK to analyze.

The Brain on PK

For each EEG recording during the PK experiment, I painstakingly edited the file to only include the time segments during which there was some kind of movement of the pinwheel. This streamlined file was then compared to Caroline's baseline EEG before performing PK. Darker colors indicate increased activity during PK, while lighter colors indicate a decrease of activity.

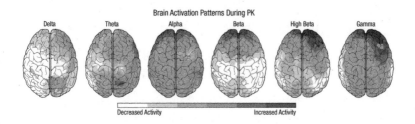

Brain Activation Patterns During PK

Delta   Theta   Alpha   Beta   High Beta   Gamma

Decreased Activity   Increased Activity

Caroline's brain showed a significant increase of fast brain wave activity (beta, high beta, and gamma) in the prefrontal lobes with an emphasis on the right prefrontal. This is similar to what we saw with Elisa (see Chapter 5) during a telepathy task. Perhaps this reflects more of a single-pointed attention. This interpretation would be consistent with what Caroline described when the pinwheel didn't move initially, and she stated that she was having trouble focusing. It is also consistent with one of the PK strategies for success identified by Dean Radin, that is, single-pointed concentration. Caroline's brain also showed a significant decrease of delta activity in the left hemisphere. A similar pattern was also observed in the theta band, where the left hemisphere showed a decrease of activity at the same time the right hemisphere showed an increase. Given what we have seen throughout this book, a left/right shift in hemispheric activation should not be surprising. Because the increased activity on the right side was seen with slow brain waves, this might suggest tapping into the subconscious. There was also a significant decrease of beta activity in the parietal region, which is primarily associated with integrating sensory information. Perhaps PK requires a shift away from the normal way of engaging with sensory information. It seems to be more about a "felt sense" of the object rather than an awareness of the specific sense qualities of the object. We have heard from several experienced PK practitioners that resonance or coherence with the object is critical. Perhaps that is what this EEG pattern reflects—a more wholistic connection beyond the normal sensory experiences.

I had a feeling there may be some additional nuance to the data that would be interesting. Using an analysis algorithm called sLORETA, I was able to look below the surface of the brain to specific brain structures. The giant spreadsheet of information that is produced shows the change in brain wave activity for every brain wave band (delta, theta, etc.) in each of seventy-nine brain regions. As I sifted through the data looking for anything that seemed unusual, one brain region stood out: the fusiform gyrus. What grabbed my attention was the fact that activity in this region decreased significantly across all brain wave bands during the PK experiment. The darkest areas in the images below indicate areas that were not assessed in this analysis. Within the specific region of interest (fusiform gyrus), darker shading in the images below indicates an increase of activity while lighter colors represent a decrease of activity. Note: These images are from the *underside* of the brain; consequently, the left and right hemispheres are reversed.

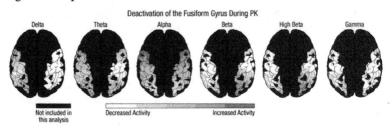

pattern reflects—a more wholistic connection beyond the normal
This part of the brain is typically involved with visual processing tasks, such as facial and body recognition. When this area of the brain is behaving differently than expected, it can lead to the experience of *synesthesia,* or the combining of normally unrelated senses. Someone with synesthesia might perceive letters or numbers as

associated with specific colors, for example. In other cases, a synesthete may perceive numbers, months of the year, or days of the week as belonging to a particular three-dimensional spatial map.

While we don't really know what it means that Caroline's fusiform gyrus became silent, we can speculate that it may be related to the ability to "see beyond" the usual and expected sensory relationships in the external world. Perhaps during a successful PK trial, the person is connecting with the overall nature of the object rather than its sensory patterns. Remember that Bannikov described the process of PK as "sensing into the energy field of the object you are trying to move"—-tuning in to its vibration, dynamics, or aura. When I shared these results with Caroline, she indicated that they made perfect sense. When she "taps in" through remote viewing, a mediumship reading, or PK, she described receiving all of the information as a singular experience. The colors, sound, and sensations are all combined into a "felt sense" related to the information coming through.

Try It Yourself

One of the best (and most common) ways to begin practicing PK is by working with an aluminum pinwheel as described throughout this chapter. This type of object is preferred as it is lightweight and super responsive. While the physical characteristics of the object should not matter (remember Luke Skywalker and Yoda), it does play a significant psychological role. If the logical part of your brain perceives the task as too hard or impossible, you are less likely to see a response.

The Setup

In general, I recommend cutting a small piece of aluminum foil (about 4 inches long and 2.5 inches wide) in the shape of a rectangle. Fold the foil lengthwise, creating a crease down the middle. Next create a stand. I like using a long thin nail pushed through a rubber eraser—this makes a solid base. Realistically, you can use anything that will support the piece of aluminum foil with minimal resistance. Unfold the aluminum foil slightly, so it forms a tent shape. Gently place the aluminum foil on the nail, centering it so that it is perfectly balanced.

The Practice

Place the tinfoil and stand on a table or desk that is a comfortable height. Be careful when placing the aluminum foil to avoid poking a hole or creating an indention with the nail. Spend a few minutes clearing your mind and becoming centered. This might involve some simple breathwork exercises to balance the nervous system (alternate nostril breathing), some simple qigong, or sitting quietly to allow the chatter of the left brain to settle down. When you feel ready, slowly bring your hands up and place them on opposite sides of the pinwheel. The movement of your hands will create an air current which will move the pinwheel. Notice what that movement looks and feels like so you can differentiate it from the look and feel of PK. Wait for the pinwheel to stop moving and simply "hold space" around the pinwheel. While you will naturally have an intention for the pinwheel to move, it is important to hold this intention as lightly as possible. See if you can let go of any effort, pressure, or expectation. Embody a positive, neutral attitude toward the foil. Trust that the foil will move and then allow it to do so in its own way, in its own time. Now, you wait. Sometimes the foil will move right away, sometimes it takes a while to "warm up," and sometimes it doesn't move at all. Movements are sometimes a slight wobble or a partial turn in one direction. Other times the pinwheel will make several full rotations in a circle or move partway in one direction, stop, and then move in the opposite direction. Treat it like a meditation. The point is to practice, not to achieve a specific outcome. When the mind becomes distracted by doubts or distracting thoughts, recognize this process at play and attempt to return to your initial orientation.

Helpful Tips

1. **Length of Practice:** While some PK teachers will recommend practicing for twenty to thirty minutes at a time, I have found that shorter practices often work better for my brain and personality. I will often practice for five or ten minutes and if nothing is happening, I let it go and move on to my other meditations.

2. **Frequency of Practice:** Regular practice seems to be helpful. While it may not be necessary to practice every day, consistency is rewarded. While the pinwheel may do nothing at the beginning, at some point it will start to move. Once this happens, the process seems to gain momentum with additional practice, becoming more automatic and reliable.

3. **External Influences:** Because the beginning practice involves working with the pinwheel in an exposed environment, it will be easily influenced by air currents in the room (including your breathing). As much as possible, attempt to find a space that minimizes drafts from air conditioning/heating, or an open window. Be careful with the movement of your hands as this will also create slight air currents. If you are concerned that your breathing is creating air currents, you can wear a face mask to minimize this impact. While these external influences can be annoying and add doubt to any movements you observe, it can also be a helpful learning tool. With time, it is possible to distinguish between movement that is created artificially and movement that is PK.

4. **Patience:** Working with PK is a serious practice in patience. Nearly everyone seems to be able to access this ability, however, it sometimes requires perseverance. If the practice is approached as an opportunity to explore specific states of attention and intention, then each session becomes a lesson.

5. **The Next Level:** The practice described above is for beginners. After you have some consistent success with moving the pinwheel, then it is time to increase the difficulty. Try placing a clear glass vase over the pinwheel or remove your hands from the sides of the pinwheel, or move farther away. By adding elements to the practice that reduce the likelihood of external influences it increases the psychological challenge. Most people find that they struggle to get movement when they increase the challenge. With practice, movement begins to happen just as it did at the beginning (refer back to tips one through four). I also recommend checking out more formal PK training programs, such as those offered by Sean McNamara, Trebor Seven, and Caroline Cory (see the Resources section).

CHAPTER TEN

ENERGY HEALING

SOMEWHERE IN MY EARLY EXPLORATIONS with medita-
tion and martial arts I heard about an ancient Chinese practice
that involved movement, breathing, intention, and visualization to
strengthen and direct the life force. Qigong (sometimes spelled chi
gong or chi kung) is an umbrella term for meditation practices that
often involve gentle movement, visualization, and awareness/circu-
lation of subtle energy (qi). These practices are a part of Traditional
Chinese Medicine and can be used for healing, spiritual development, and/or martial arts (such as tai chi).

At the time, the ideas and concepts behind qigong were completely new to me and seemed somewhat naïve or simplistic. I grew
up in the Midwest where nobody was talking about these concepts
or practices. In fact, at that time and place, meditation and yoga
were still considered "out there" and "dangerous." Despite my conditioning, these energy practices made intuitive sense to me. They felt
right. I was intrigued and wanted to learn more. Through tenacity,

luck, serendipity, synchronicity, or some combination of all of these I stumbled upon a small group of qigong practitioners that met on Sunday mornings at a nearby interfaith church led by Greg McDonald.

Greg was a PhD-level biochemist. He left a university position to pursue other interests, including teaching qigong. He had gathered practices from a variety of traditions and blended them together into his own style, which included elements of Shamanism, Native American spirituality, Taoism, Buddhism, and New Age practices. Having no prior experience, I assumed that what Greg was teaching was traditional qigong. I later learned that there are literally thousands of types of qigong, making it a broad and diverse set of practices.

Greg was more interested in helping each of us cultivate the ability to feel and sense energy rather than rigidly focusing on the correct way to do the movements. He assumed that each of us could sense and manipulate energy fields for the betterment of others. This confidence was valuable for me as it allowed me the space to trust my feelings and any intuitions I might receive. No matter what I felt, or experienced, it was all considered valid.

In any given class we might experiment with "passing qi" to each other or tapping into the healing potential of the Earth, trees, stars, or each other. It was here that I first learned to feel energy and began experimenting with energy healing.

As I became more sensitized to the feeling of energy moving in and around my body, I became slightly obsessed. This was amazing, and I wondered why more people didn't know about this. I had discovered that there was a concrete way to directly experience something beyond the physical self. It was firsthand evidence that I was more than a pile of organs in a "meat suit." Within a few years I was teaching my own qigong classes and developing my own way of understanding how this all worked. Greg's confidence in my ability and his encouragement to create my own style helped me to realize that there isn't a single path to energy development. In fact, it felt important that each of us find our own way. While this general attitude resonated deeply within me, it also seemed contrary to what I had observed. Most energy healing schools and systems seemed to suggest that they had the only valid approach. Instead, Greg taught me to explore, experiment, and trust my Self.

The Healing Group

A subset of students from our qigong group started meeting outside of class to practice energy healing techniques. Typically, we would work out of a simple room in a rented space or in someone's home. The healing space would basically consist of a few chairs, a massage table in the center, and whatever "healing objects" the members contributed. One of us (or someone we knew that needed healing) would lie on the table and our group would do their thing. Not surprisingly, it was an eclectic group and set of approaches. Some of the healers would sit back in their chair with their eyes closed, others would approach the person on the table and put their hands on the area that needed healing. Some people would move their hands around the person to help clear the energy, others used feathers, crystals, or other tools. Essentially, each of us was attempting to listen to our inner guidance and do what felt right. Nearly every time, the person receiving healing felt some significant relief during and after the session. Sometimes they would describe experiencing a sensation of energy, like electricity, moving through their body, or drifting into a trance-like state, or seeing vivid colors. Of course, it was impossible to know if the effects of the healing had anything to do with the intention of the healers in the room or if it was simply placebo. We know from decades of research that if you truly believe something to be true, it can be made manifest in the physical body. Maybe the people being healed were simply putting their faith in us and that was enough to make it work. Although sometimes the experience of the healer and the person being healed were so synchronous there certainly seemed to be a connection. There were frequently times that someone would be at the center of a healing and afterwards would describe images or sensations they received during the session, which were nearly identical to the intentions of one or more of the healers. For example, in one circle, I was visualizing sending "Earth energy" toward the person for healing. After the session, the person being healed described feeling "extremely grounded," with images during their session of "being in a clearing in a forest, surrounded by roots that were supporting them." Another time, one of the group leaders, Nellie, was working on a woman who had intense pain in her upper back. Nellie tuned in and felt a heaviness in that area, a stagnant energy. Using her intention, visualization, and hand movements, Nellie imagined scooping this energy out and releasing it back to the Universe. At the same time that Nellie sensed this energy beginning to move, the client let out a heavy sigh and described having significant relief. Stories like these could certainly be coincidence, but when you see things like this repeatedly, coincidence seems less and less likely.

Energy Healing Research

During the time of my qigong energy-healing exploration, I started looking into research with energy healing. I was pleasantly surprised to find that there were a significant number of studies demonstrating the effectiveness of practices such as prayer, reiki, and therapeutic touch in reducing symptoms for a wide range of concerns. Unfortunately, a lot of this research suffered from study design limitations that made it difficult to know if the positive effects were due to the energy healing, attention from the healer, a belief in the power of healing (placebo effect), or other unspecified lifestyle changes. Because of these concerns, a few researchers have conducted gold standard studies, randomizing subjects into active healing and control groups, blinding both the researchers and the participants to which condition they were in, and controlling for a range of participant variables including health condition, age, comorbidities, and so forth.

In addition, many of these sophisticated studies utilized distance healing, rather than hands-on or in-person healing, to further eliminate possible confounds.

One of the first of these studies examined the impact of prayer on 393 coronary-care unit patients. When participants were admitted to the hospital, they were assigned to either a prayer group or a control group. Because it was a blind study, the patients did not know if they were being prayed for or not, eliminating the possibility of the placebo effect. Those in the prayer group were matched with three to seven intercessors who were given the patient's name, diagnosis, and general condition, as well as updates on their condition throughout the study. The intercessors were from a variety of Protestant and Roman Catholic churches and were asked to pray daily for the "rapid recovery, and for preventions of complications and death, in addition to other areas of prayer they believed to be beneficial to the patient" (Byrd, 1988). After analyzing the results, they found that the prayed-for group had significantly fewer cases of pneumonia, congestive heart failure, intubation/ventilation procedures, cardiopulmonary arrest, and significantly less need for antibiotics and diuretics. The prayed-for group was also more likely to demonstrate a "good" hospital course.

Another well-done study examined the impact of distance healing on people with advanced AIDS (Sicher et al., 1998). Forty participants were matched based on their age, T-cell count, and number of AIDS-related symptoms. These matched groups were then assigned to either a distance-healing group or a control group. Distance-healing practitioners from a variety of traditions served as the healers and sent healing energy to the identified patients for one hour each day for six days. Each patient in the active distance-healing group received healing from ten different practitioners. During the six months of the study, patients in the treatment group experienced significantly fewer doctor visits, hospitalizations, and new AIDS-related diseases. They also had shorter periods of hospitalization, lower severity of illness, and improved mood compared to the control group. There were no differences, however, between groups in physical symptoms or quality of life.

Beyond these two examples, summaries of well-designed, energy-healing research have been conducted by a few research teams. Astin and colleagues (2000) examined the results of twenty-three studies that met specific criteria, including random assignment of participants, a placebo control group, and publication of the results in a peer-reviewed journal. Overall, the improvement in conditions treated was statistically significant with a combined effect size of .40 (p < .001). This means that there was less than a 1 percent chance that these results were due to chance and a 99 percent chance that these findings reflect a valid finding. Of the twenty-three studies examined, 57 percent (thirteen) showed a positive treatment effect, nine showed no effect, and one showed a negative effect. While this may seem a bit disappointing, it makes sense that you would not see positive change in every study. Despite best attempts, there are many factors that cannot be controlled in research. What if the control group just so happened to be more physically active, or have a better diet, or a better support system? Or what if the control group also had friends and relatives praying for them that nobody knew about? The fact that more than half of the studies reviewed showed a positive impact from the energy healing despite these unknown influences was enough for the researchers to suggest that this area deserved more attention.

A more recent meta-analysis of distant healing studies was done that involved "nonwhole human target systems." This means the healers were sending healing intentions to animals, tissue samples, or bacteria. The reason for this approach was to create even more control of the study, creating greater confidence in any of the findings. A tissue sample, for example, does not have any beliefs about healing that could interfere with the results, and it is much easier to control the environment. After combining the results of forty-nine studies from thirty-four research papers, the authors found a small, but statistically significant effect for the impact of distance healing (Roe et al., 2015).

This type of research was important and probably necessary for me to fully accept that there is more to energy healing than wishful thinking or a placebo effect. For me, the logic was straightforward. If humans can influence cells with their intention, they should be able to influence the things made of cells (organs, organ systems, and whole organisms).

Despite the rather convincing research evidence showing that our thoughts can influence other systems, there was little research exploring what is happening in the brain during energy healing, either for the "sender" or "receiver."

Grand Master Ken Cohen

Ken Cohen has studied and taught qigong, tai chi chuan, and Taoism for more than fifty years, becoming one of the world's foremost authorities on these ancient traditions. He is the author of the book, *The Way of Qigong: The Art and Science of Energy Healing*, as well as more than two hundred articles on related topics. Near the beginning of my qigong journey, Ken's book provided an important part of my education, offering a scholarly approach to the history, science, and applications of energy work from a qigong perspective.

Eventually, I enrolled in his Qigong Instructor Training Program, allowing me to learn from him directly. After several years of workshops, private lessons, book reviews (and a few tea ceremonies), I invited Ken to teach a Primordial Qigong workshop in Columbia, Missouri. At the time, I was working at the University of Missouri as a health psychologist and found some additional support from a Contemplative Practice Initiative on campus. Having Ken in town for an entire week seemed like the perfect opportunity for a case study. I mean, how often do you get the chance to conduct an EEG recording of a Qigong Grand Master?

As the week progressed, I managed to find a window where Ken was open for a few hours. I asked him if he would be willing to participate in a brain wave study while engaging in energy healing and meditation. Turns out, he had done some EEG brain mapping research many years earlier and was eager to find out how the two sessions compared. I had no idea at the time, but Ken had been one of the research subjects in the famous Copper Wall Experiment conducted by Elmer Green, PhD, between 1983 and 1995 (Green, 1991).

Green was a pioneer in the field of biofeedback and consciousness. To scientifically measure energy healing, he created a room with a copper wall that was wired up to record changes in electrical activity. Green then brought in a variety of energy healers (including Ken Cohen) to sit in the room, meditate, and send healing energy through intention. When Green analyzed the data, he found that the wall was picking up massive surges of electricity that would come in bursts, usually lasting only four to five seconds, but at much higher amplitudes than would be expected from random fluctuations in the body. In fact, Green reported on one occasion that he recorded a reading at 200 volts, which is 1,000 times larger than would be expected from electricity generated by the body. In many cases, these bursts happened at the same time that the healer was generating or sending a healing intention. Other times, the surges just occurred while the healer was engaged in meditation. I didn't have a copper wall, but I did have EEG equipment.

The Healing Session

For our case study, we invited one of the participants in the weekend workshop to serve as the client. Because this wasn't a formal session, we decided it was best to do a general wellness protocol rather than attempting to treat any specific concerns or conditions. After attaching an EEG cap and recording some baseline data, Ken stood near the volunteer with his hands above her head, facing the crown chakra. He slowly and simultaneously brought his hands down the front and back of the body, as if he were feeling and sensing her energy field. At times his hands would move in slightly toward the body, and at other times they would push out away from her body. Once his hands reached the height of the waist, he would release his hands and return to the top, repeating the process a few times. Later, he indicated that this aspect of the protocol was designed to balance the two primary meridians in the body as well as the chakras. After repeating the balancing process a few times, he moved to the front, facing the volunteer, with one hand pointing toward the Earth and the other sweeping up and down about eighteen inches away from her body. Again, after a few minutes, Ken repeated the same sweeping process, but facing the back of the volunteer. He indicated that this part of the protocol was designed to smooth the energy in all of the meridians. For the final part of the protocol, Ken asked the volunteer to lie down on the massage table. He began by sweeping both hands down the energy field of the body with the intention of removing any stale, stagnant, impure, or toxic energies that may be present. He continued this process until he had the sense that the system was as clear as it was going to get. The next step was to provide the body with additional nourishing energy for self-healing. He did this by spending a few minutes with his hands over each of the five yin organs (liver, heart, spleen, lungs, kidneys). He would place his hands slightly above the organ he was focusing on and imagine sending specific colors of healing light to each one.

A Qigong Healer's Brain

Ken's brain wave activity was measured first in a baseline condition, simply standing with his eyes open doing nothing. Next, his brain wave activity was recorded during the healing session.

After the data were cleaned to remove any artifacts from movement, the two recordings were compared.

The brain images below show the percent of change in each brain wave (delta—high gamma) comparing healing with baseline. The key below each head map indicates the amount of change. For example, the graph under the gamma head map ranges from -73 to +73. This means that anywhere the gamma head map shows the darkest shading there is a 73 percent increase in brain wave activity in those regions. Areas with the lightest shading indicate a 73 percent decrease in activity.

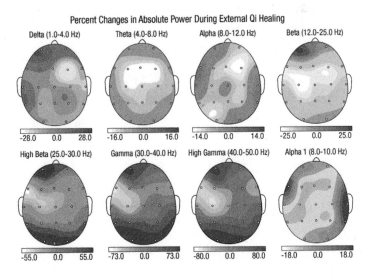

Percent Changes in Absolute Power During External Qi Healing

Clearly, the largest changes that occurred during an external qi healing related to increases in fast brain wave activity (high beta, gamma, and high gamma) in left prefrontal and occipital regions. Interestingly, increased gamma in the left prefrontal regions has been seen in Tibetan monks engaged in a lovingkindness/compassion meditation (Lutz et al., 2004). It is also associated with an approach orientation toward others and a positive outlook (Kelley et al., 2017). The strong increase in occipital activation suggests visual processing. These areas may be involved while the healer "sees" the areas of the energy body in need of healing or visualizing healing energy moving into the body. These findings were intriguing and seemed consistent with what I knew about energy healing. I wondered if similar patterns would show up in healers that did not have experience with qigong. Shortly after the experiment with Ken Cohen, I had the opportunity to find out.

I was in Massachusetts coteaching a mediumship workshop with Joanne Gerber. My role was to explain some of the brain imaging findings to the audience while Joanne led the group through a variety of meditations and mediumship development exercises. I offered private brain mapping sessions to any of the participants so they could have data related to their own processes. While most of those signing up for these sessions wanted to see what their brain was doing when they were tuning in to their mediumship abilities, one of the participants was more interested in what was happening during energy healing. Katherine Glass was a professional psychic who offered energy healing sessions through her business, the Healing Essence Center in Concord, Massachusetts. Katherine had been trained in both Brennan Healing Science and Reconnective Healing. During our session together, this became the primary focus. She was curious about what was happening during a healing state, and I was curious if her brain patterns would look anything like Ken Cohen's.

Because we didn't have another subject for the healing session, we decided that Katherine would work on me. I had some minor pain in my right knee related to a martial arts injury. This provided a good target for her work. Katherine put her hands on either side of my knee and seemed to enter a kind of meditative state. She didn't say anything and barely moved, making it easy to collect good EEG data. I was simply trying to be open and receptive, while also keeping one eye on my computer. After about five minutes or so, Katherine indicated that she was finished. I stopped the EEG recording, made a few notes, and then recognized that the pain in my knee was gone. It wasn't that bad to begin with, maybe a three or four on a ten-point pain scale, but now there was nothing. This certainly caught my attention as I am generally not overly susceptible to suggestion. If anything, I tend to "get in the way" of receiving readings or heal-ings because I am too busy thinking about what is happening and attempting to be objective.

Later that night, I analyzed Katherine's data, comparing her base-line EEG to the EEG from the healing session using the same system I used to examine Ken Cohen's data.

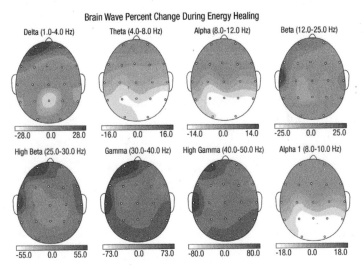

Brain Wave Percent Change During Energy Healing

The similarities were remarkable. During the healing session, Katherine showed a significant increase of fast brain waves, just like Ken Cohen, with an emphasis in the same areas—the left frontal and occipital regions. Katherine also showed some fast activity near the ears, which is likely due to jaw tension rather than brain waves. The other striking feature was a significant decrease of slow wave activity (theta and alpha). This was particularly interesting given that these recordings were conducted with eyes closed, when we expect these brain waves to be the strongest. This significant decrease of alpha, focused at the back of the brain, may indicate an external focus. As mentioned in Chapter 8, increases of alpha tend to correlate with an internal attention while decreased alpha suggests an external attention. Even though Katherine's eyes were closed, she was focused on me and my knee (external).

It seemed curious that two different energy healers with completely different backgrounds and training demonstrated near identical brain wave patterns. This made me wonder if there were some common elements to many/most healing traditions related to how they were using their attention and intention. Perhaps that is what was being revealed in these case studies.

In the book, *The Power of Eight: Harnessing the Miraculous Energies of a Small Group to Heal Others, Your Life, and the World,* award-winning author and researcher Lynne McTaggart describes the basics of an effective healing approach. While her program is oriented toward small groups, I suspect that the same fundamental processes apply to individual healing sessions. The keys to success, based on research evidence, include: believing in the process, focusing the mind, quieting the mind, connecting to the object of healing, visualizing the outcome, mentally rehearsing, and letting go (2017). This description is consistent with what was observed in the brain wave patterns. In particular, the intentions related to connecting to the person being healed and visualizing the outcome appear to be directly related to the brain regions that were activated in these case studies.

Pranic Healing (PH)

At the University of Missouri, I taught Mindfulness-Based Stress Reduction (MBSR). Our class would meet for two hours each week for eight weeks, covering specific topics related to mindfulness and then practicing various forms of meditation together. During one of these meditation sessions, I noticed a student sitting on her zafu (meditation cushion) making strange movements with her hands. After watching her for a few moments, I noticed a pattern. First, she would move her hands toward the center line of her body, as if she were pressing against an invisible wall. After doing this a few times, she would then move her hand in front of her body in a circular motion and then make a flicking gesture to her side. This process lasted for thirty seconds or a minute and then she would repeat the process. While curious, I was also slightly annoyed. Whatever she was doing, it was not the meditation I had just introduced. Eventually, I asked this student what she was doing. Her name was Neus Raines. She was originally from Spain and was completing her PhD in economics in the United States. She told me that she had been training in pranic healing and arhatic yoga for the past several years and was doing a self-healing practice during the meditation session. I had never heard of Pranic Healing or Arhatic Yoga and honestly had no interest in either of them. At the time, I felt like my mindfulness and qigong practices were all I needed. Everything else was a distraction.

Neus discovered that I was exploring EEG patterns associated with meditation and suggested we look at pranic healing. I declined. A few months later, she followed up, providing me with a bit more information and some ideas for a simple study. Again, I declined; still not interested. Being tenacious, Neus waited another few months and contacted me again, assuring me that the study would be very simple, and my involvement would be minimal. For whatever reason, this time I decided to listen. I'm glad I did—our little study ended up having an important influence on my personal and professional explorations.

Having no idea what pranic healing was, I had to do some research. I knew what healing was, but wasn't clear about the term pranic, I thought maybe it had something to do with the breath. Neus let me borrow her book, *Miracles Through Pranic Healing*, by Master Choa Kok Sui to help me get oriented. Turns out, the root of pranic is prana, which is simply another term for life force or qi. In essence, pranic healing is a no-touch energy healing modality based on the notion that it is possible through intention and simple hand movements to manipulate the prana in someone else. If you can detect imbalances in the other person's pranic field and help remove stagnant energy and fortify depleted areas of the energy body, you can help restore health and well-being. This was apparently what Neus had been doing during our MBSR class. She was first "testing" or sensing the state of her energy in each chakra and then using circular hand movements and intention to either remove stale energy or increase the amount of healthy energy in that area. After clearing an area, she would "flick" the unwanted energy away, imagining that it was being consumed by a green fire at her side.

Neus's idea for the study was to see if there were any significant brain wave changes that occurred when someone was receiving a pranic healing treatment session. This was interesting because the previous case studies I had conducted were looking at the healer. As far as I was aware, nobody had investigated whether the person being healed showed any changes in their brain waves.

Of course, there was still the ongoing challenge of ruling out or controlling for the role of belief or placebo. I learned from my energy-healing literature review that belief and intentions are powerful. If someone with pain is willing to come in for a session with an energy healer and then goes through a sixty-minute healing session, it is likely that they may feel better simply because they believe they will feel better, or perhaps because they received caring attention from someone. Because we are a mind/body, if we think we are better, the brain is likely to respond accordingly. Our solution was to have each participant come to the clinic twice, one week apart. During one session, they would receive a pranic healing treatment and during the other, we would fake a healing session. Because pranic healing does not involve touch, we created a testing situation in which the healer (Neus) would stand approximately four feet behind the subject who was seated in a chair. The subject had no way to see what Neus was doing. To minimize the likelihood that the subjects might pick up on subtle cues and figure out which of the sessions involved actual healing, Neus attempted to mimic some of the natural movements that might happen during a typical session during the control condition. For example, she would shift her weight, circle her hands, and even use a spray bottle containing alcohol and lavender essential oils to cleanse her hands periodically. Even though we only had four subjects, we counter-balanced the order in which each person received the conditions. Two of the subjects received the control condition first, while the other two received the pranic healing session first. At the end of the study, we asked each participant to guess which of the sessions was the actual pranic healing session. Half of them guessed correctly, which is what you would expect by chance and suggests that there were no obvious overt cues.

At the beginning of each session the participants rated their level of pain, discomfort, tension, and mobility, then we measured their

EEG activity using a nineteen-channel system. After their session, we repeated these measurements. This process allowed us to compare the results of the actual pranic healing session to the placebo/control condition. The results were intriguing.

On the subjective measures, three of the four subjects reported about the same change in pain during the two conditions. However, Subject Four reported an 83 percent decrease in pain after the PH session and only a 17 percent decrease in pain after the placebo condition. All subjects reported a decrease of 50 to 83 percent in discomfort following the PH sessions, while the placebo condition resulted in discomfort reductions between 0 and 50 percent. For tension, Subjects One, Two, and Three reported similar reductions in both sessions (50 percent) while Subject Four reported an 83 percent reduction following PH and a 0 percent reduction after the placebo. There were no significant changes between sessions in mobility. While these results weren't definitive proof, they seemed to favor the PH sessions, suggesting there is something more happening above and beyond placebo or relaxation. It was also interesting that each subject seemed to respond somewhat differently. For example, Subject Four reported a much more dramatic shift during PH than during placebo on several variables, while most of the other subjects showed more minimal differences.

Pranic Healing and the Brain

We examined changes in the brain by comparing each participant's presession data to their postsession data. This gave us a picture of how each session changed brain wave patterns, allowing us to see the impact of both the placebo and the PH treatment. Every way we sliced and diced the data, the results showed that the brain changes occurring during PH sessions were significantly different than those occurring during the placebo condition. Interestingly, none of the subjects showed the same *kind* of changes. For example, Subject One showed a significant increase of slow brain wave activity (darker shading) during the placebo condition, but a decrease of the same brain waves during the PH session. Subject Four showed a significant increase of alpha activity during the placebo condition, but not during PH. This participant also demonstrated a significant *increase* of gamma activity in the back of the head during the PH condition but not during the placebo.

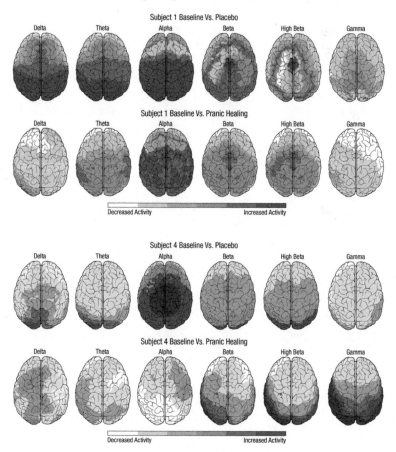

Interpreting the Data

Even though there were significant differences between the placebo and PH sessions for each subject, there did not appear to be any consistent patterns between subjects—everyone seemed to respond differently. Despite some initial frustration, these results made perfect sense. Why would I expect a group of distinct individuals with different ages, gender identities, backgrounds, education levels, incomes, health conditions, character traits, and temperaments to respond the same to an energy healing session? It makes more sense that everyone would respond in a personalized way. It also makes sense that some people might have more dramatic effects than others. Of course, studies with a small sample size tend to raise more questions than they answer. For example, are the brain wave changes during a healing session different from what is observed immediately after a session? What about two weeks later? Are the brain wave changes demonstrated by Subject Four more consistent with greater subjective changes? This person clearly reported the largest reduction in pain and discomfort following the PH session compared to the other participants. Is increased gamma activity the key, or can any of the brain waves be associated with healing? I also wondered if the "openness" or receptiveness of the person being healed might also make a difference. What if Subject Four got the most from the PH session because they were the most open and resonant to the healer or the process? Clearly, we still have a lot to learn.

Try It Yourself

The ability to heal ourselves and others appears to be innate; a capacity we each possess. But there are a variety of strategies, techniques, and states of consciousness that can facilitate the effectiveness of this ability. In addition to the suggestions offered earlier from Lynne McTaggart's work, it also seems important to dis-identify from the healing process. Most successful healers do not view themselves as causing the healing. Rather than believing that they are using their own personal healing energy to cause a response in the other person, they believe they are acting as a channel for the healing energy of the Universe. This mindset seems important in a few ways. First, it removes ego from the situation. If it is not really about the healer's power, it becomes easier to get out of the way and allow the healing to happen on its own. Second, this mindset implies that the healer trusts the process and the healing capacity of the other person. It can be empowering to realize that you have the capacity to heal yourself. Third, it minimizes the tendency to over-effort. When you attempt to enforce your intention through willpower, the resulting tension often blocks the flow of energy that naturally happens. In fact, one of the best things you can do as a healer is stay relaxed. This "healer as channel" approach can be practiced through the process of holding space.

Holding Space

"Holding space" is a concept that has been popularized in ceremonial gatherings and psychedelic experiences to describe the primary role of a sitter.* These are people with experience and/or training who help create an energetic container with their presence. Their job is primarily to maintain a grounded, calm, steady, and compassionate disposition. The presence of one or more sitters in

*The term sitter as used in this context is short for "trip sitter," and borrowed from the psychedelic world. The term sitter, when used in the context of mediumship and/or psychic readings, refers to the person receiving the reading.

an intense experience can be very stabilizing for the participants, offering emotional and physical safety and support so that participants can explore aspects of themselves, express difficult emotions, and heal.

In many ways, this is the same approach used by successful healers: creating a safe and supportive space, setting a clear intention, and trusting that the other person can heal themselves. With this in mind, being a sitter or holding space is like a meditation: letting go of personal thoughts and concerns, being aware of what is happening internally and externally in the moment, and maintaining a gentle, positive regard for the other person(s) present. When I was completing my training as a Psychedelic Guide through the Medicinal Mindfulness program (working with cannabis as a psychedelic medicine), they explained this sitter orientation as a 60/40 division of attention. Sixty percent of your attention is on yourself—being aware of your thoughts, feelings, and reactions, quieting your mind, noticing the energy in the environment and its impact on you, staying relaxed, and breathing in a rhythmic gentle fashion. Forty percent of your attention is on the person or group: tracking their reactions, offering support if needed, sending positive intentions. I love this approach as it offers a method to check your own process. As you are working with someone, either as a sitter or a healer, how is your attention divided? Are you pushing? Are you trying to influence the process? Is your entire attention on the other person? Are you distracted by your own thoughts? Periodically, while holding space, you can simply check in with yourself and recognize where that percentage of attention is at the moment and if it requires an adjustment on your part. You can use this same mindful awareness throughout your day to practice this skill. While you are in a meeting, where is your attention? What about during a conversation? How about when you are checking out at the grocery store? While these different situations may require slightly different percentages of attention, the fact that you are becoming aware of this process is the key and will make it easier to find the right balance when you are helping others in a healing role.

Healing Feedback

One of the tricks to developing healing skills is knowing if you are actually making an impact. Yes, your clients may report feeling better, and they may be willing to keep coming back and paying you money, but that is not proof that you are actually doing anything. As we discussed earlier in this chapter, the placebo effect is powerful. A good way to experiment with your impact is by experimenting with your impact on nonhuman targets. Here are a few ideas to get you started.

The Rice Experiment

This has become a relatively famous experiment originally attributed to Masaru Emoto, a Japanese businessman who studied changes in the molecular structure of water based on what intentions were sent to it. This slightly modified version of his experiment is relatively simple but can provide direct evidence of the power of intention.

1. Place the same amount of uncooked rice in two separate jars.
2. Cover the rice in each jar with the same amount of water.
3. Each day send healing intentions to jar number one (i.e., practice holding space). Do this for at least five minutes every day.
4. Ignore jar number two.
5. Continue this every day and notice what happens after several weeks of this practice. You may notice that the rice sent

healing intentions remains white, while the ignored jar begins to become discolored.

6. Continue for several months and observe how the two jars look more and more dissimilar.

The pH Experiment

We measure the acidity or basicity of a solution on a pH scale. In general, clean drinking water should have a pH of about 7, or between 6.5 and 8.5. Various researchers, including Dr. Gary Schwartz, Dr. Melinda Connor, and Lynne McTaggart, have conducted experiments to see if healers (and ordinary folks) can change the pH of a solution simply through their intention. The answer, as you might have guessed, is *yes*. For this experiment, you will need a pH meter, which can be purchased at most hardware or lawn and garden stores.

1. Obtain a baseline pH reading of a glass of water obtained from your sink, noting any fluctuations over a several minute period.

2. Practice "holding space" and sending gentle but focused intentions to the water to either increase or decrease the pH level.

3. Keep practicing. If you don't notice anything the first few times, try making some adjustments to your approach and don't be surprised if the pH shifts in the opposite direction of your intention; sometimes that happens.

The Seed Experiment

This experiment is basically the same as the rice experiment except rather than focusing on rice, you are directing your attention toward one group of seeds while the other group is ignored.

1. Get your seeds. Previous research in this area has used barley or wheat seeds, but I suspect just about anything would work.

2. Create two identical pots (labeled number one and two). Use the same amount of soil from the same source.

3. Plant a dozen or more seeds in each pot, making sure you plant the same number of seeds in each one following the instructions on the seed packet.

4. Ask a friend to randomly choose one of the pots to receive loving intention.

5. Keep the pots in the same location.

6. Every day for five minutes, practice holding space with the identified pot of seeds, sending it positive, loving regard. During the intention sending, separate the pots to avoid inadvertently sending positive intentions to the control seeds. After the five-minute practice, return both pots to their original placement.

7. Continue this practice for a few weeks and then measure the length of each seedling in the two pots. Average the seedling height in each pot and see if there is any difference.

CHAPTER ELEVEN

LESSONS LEARNED

THROUGHOUT THIS BOOK we have seen that certain brain waves and brain regions are consistently associated with psi abilities. Sometimes these patterns seem specific to a particular ability and other times they seem to cross over. While not everyone shows the same activation or deactivation patterns, the fact that the same brain regions and brain waves keep showing up suggests that there is something important about their functioning in relation to psi. In fact, these brain patterns seem to be directly related to certain attitudes, intentions, and states of consciousness. When we practice and develop these mental states, we change the brain. When we change the brain, we create an opportunity to develop these skills. In this regard, an ideal psychic development approach might involve both the training of specific mental states and the creation of ideal brain conditions.

The Five Mental States

Most of the psychics, mediums, and healers studied describe five mental states necessary for successful psi. These can be summarized as focus, connection, staying relaxed, listening, and trusting.

Focus. This refers to how the psi practitioner is directing their attention toward the task at hand. In some cases, it seems to require a single-pointed concentration, while at other times it requires a broader attentional state, becoming aware of any information that moves into consciousness without screening or directing.

Connecting. Whether connecting to another person such as might occur with telepathy or energy healing, connecting with a disembodied entity such as a spirit guide or discarnate human, or connecting to a specific object such as in ESP games or PK, many practitioners describe a sense of moving beyond their limited, normal ego boundaries to establish a sense of empathy or resonance with the other.

Relax. This can take many forms, including physical relaxation (energy healing, PK), but also mental relaxation, which of course is connected. In general, if there is an excess amount of mental efforting involved in a task, it is less likely to occur. Practitioners describe more success when their practice is in a state of flow, happening easily and automatically. When we overthink or attempt to force these skills, it often doesn't work.

Listen. In this context, listening means listening to the soft, subtle voice within. Whether this is considered intuition, your Higher Self, or a disembodied guide or helper, there is a different kind of attention required to hear (or see or know) this information. It is often quiet and generally drowned out by the louder, more insistent conscious mind.

Trust. This relates to the ability to accept the reality and accuracy of the information received when listening. If we second-guess information received, or get caught in doubt or analysis, it corrupts the accuracy of the information and leads us astray. Of course, it is still important to be discerning and recognize when the conscious mind is taking over, but when we truly "get a hit" it requires acknowledgment. When we learn to truly trust the information received, we tend to get more.

These five attitudinal qualities of psi activity seem to be reflected in the brain changes observed in our studies. In fact, we can use the brain imaging data to help clarify and expand our understanding of the mental skills required for psi. Below is a simplified summary of our findings.

The Brain on Psi: A Summary

Right Hemisphere. In general, we have seen that there are more changes in activity in the right hemisphere than the left across all psi abilities. The left hemisphere is typically responsible for language processing. Thinking too much (which is typically done through internalized language) seems to inhibit psi abilities. By contrast, the right hemisphere is generally more involved in experiential processing—a more subconscious/automatic mode of engaging in the moment. This suggests that psi abilities are more easily accessed by shutting down the language-oriented, analytic, problem-solving, linear processes of the left hemisphere; and allowing the automatic natural processing of the right hemisphere to take over.

The Default Mode Network (DMN). The DMN is a group of brain regions that work together to facilitate certain states of consciousness. In particular, the DMN is responsible for creating our sense of self, our identity or ego. Any time we are thinking about ourselves, we are engaging the DMN. This network serves as a sort of internal structure that allows us to filter information from the environment and create meaning from what we experience. For many of us, this process limits our ability to believe or experience anything that does not fit into our societal "programming." Temporarily shutting down or disrupting the normal activity of the DMN can help increase our psychological flexibility that is associated with psi-related experiences.

The Right Parietal Lobe. This region of the brain is involved in establishing boundaries between self and others. When the normal functioning of this area is disrupted, it seems to facilitate perspective taking and the ability to connect with energies outside of the small self. Changes in this region of the brain seem to typically occur through an increase of slow brain waves (delta and theta). This pattern appears to be most related to mediumship, spirit communication, and channeling.

The Right Frontal Lobe. This area is implicated in sustaining and directing attention. At times this region seems to become activated, sometimes deactivated, and sometimes showing indications of both activation and deactivation at the same time. This suggests a change in the way the person pays attention, which should not be surprising. At times the attention may need to become more focused, and at other times it may need to become more spacious. It may be directed internally or externally, or some kind of combination of the above.

The Occipital Lobes and Gamma. The occipital lobes represent the most posterior section of the brain and are largely responsible for visual processing. This area of the brain frequently "lights up" during psychic and mediumship practices. While there are several brain wave changes observed across practitioners, there is a relatively consistent increase of gamma brain waves. Gamma is the fastest brain wave and is associated with activation, as well as higher states of consciousness, such as being in a flow state, during psychedelic experiences and lucid dreaming. It is thought that this gamma activation in the occipital lobe represents the "inner seeing" of the practitioner. While not all practitioners receive psychic information visually, it is possible that this activation represents a broader perceptual processing. Gamma activation does not seem to favor the right or left hemisphere. The area of activation may be related to which visual field the practitioner is engaging to "see" or process the psi information. If the information is processed in the left visual field, we would expect to see the right occipital lobe become engaged and vice versa.

Theta. Theta represents one of the slower brain waves, occurring between 4 and 8 cycles per second (4 to 8 Hz). With mediums and psychics, excessive theta often shows up at baseline in the frontal lobes, when they are not engaging in any particular skill, but just sitting doing nothing. When this is observed outside of psi-related activities, it tends to be associated with ADHD, that is, difficulty sustaining attention and problems inhibiting thoughts, emotions, or behaviors. It seems possible, if not probable, that certain ADHD characteristics may be helpful with psi abilities. For example, having a tendency toward disinhibition and a diffuse attention may make it easier to accept new or unusual information without screening it out. While this may make it difficult to attend to everything in the physical world, it may make it easier to tune in to the spiritual world. Supporting this notion, theta is also associated with connecting to subconscious aspects of the mind.

Given that these brain regions/brain waves seem important to psi-related abilities, the obvious question is, "How can I get my brain to do that?" Clearly, some people are just born this way. People like Laura Lynne Jackson, Joanne Gerber, and Jeannine Kim appear to have a natural ability to shift their brain wave activity in just the right way to access these abilities. But what if you are not naturally gifted in this way? Is it possible to retrain the brain, to help it become more flexible and fluid, in other words, to become more psychic?

Brain Retraining

Without fail, every professional medium, psychic, and energy healer I have interviewed has told me that everyone has these abilities; the trick is to learn how to access them. As Trebor Seven stated, "It is more like remembering that you have these abilities and learning to trust yourself." I asked Caroline Cory, director and writer of the documentary *Superhuman*, if she believed that anyone could develop psi skills. She explained that it is not really about developing but more about uncovering. She explained that children are often able to demonstrate psi abilities easily. You just ask them to do something, like read a book through a blindfold, and they do it. They don't think about it. They don't consider whether it is reasonable. They just do it, having not yet been convinced that it is impossible. Apparently, this fluid, flexible way of connecting with the world typically lasts until about the age of eight, after which it is gradually replaced by more sober, "rational," limited thinking and beliefs. Interestingly, age eight is about the time that brain waves begin shifting from a theta-dominant state toward more of an alpha-dominant state. It is also around this age that most children have fully entered the concrete-operational stage of cognitive development. Concrete-operational is the third stage of a cognitive development theory proposed by Jean Piaget (1954) and is characterized by the development of organized and rational thinking (sounds very left-brained).

Perhaps what happens is that we all begin with a strong natural ability for psi. Our brains are open, flexible, and nonjudgmental. As we get older, the brain changes. The brain waves speed up and activation moves more toward left-hemispheric dominance. We are "rewarded" for this way of processing information through formal education and cultural influences, leading to a significant rewiring of the brain and a closing down or inhibition of our natural psi abilities.

In this context, developing psi abilities is more of a remembering or an unlearning.

Apparently, we need to learn how to maintain enough cognitive and psychological flexibility to access more of our right-brained processing, while limiting the influence of the left brain. For some, this tendency may occur naturally or easily, for others, it may require significant brain (re)training. The brain training methods described below can be used to help unlock or improve psi abilities.

Meditation

Throughout this book we have discussed ways that meditation may be a helpful tool in exploring or unlocking psi potential. Of course, meditation is not a single practice. Meditation is an umbrella term that covers a wide range of practices. At the NeuroMeditation Institute we have identified five primary styles of meditation based on the way that attention is directed, the intention, what brain waves are involved, and in which brain regions. Each of these styles may offer something particular to the uncovering of psi abilities. Note: You can find guided meditations for each of the five styles on the NeuroMeditation YouTube page (@neuromeditationinstitute).

Focus Meditation

Meditative practices with this emphasis involve sustaining attention on a single object or intention. Basically, you direct attention toward a specific target—such as the breath, or a mantra, or your third eye. When the mind becomes distracted and begins searching for something more interesting, you recognize this as quickly as you can and gently shift your attention back to the target. These practices activate and exercise the frontal lobes (particularly the right frontal lobes), making it a perfect practice for anyone hoping to improve their attention, memory, or other cognitive functions. We saw several examples of the right frontal lobes becoming activated during tasks such as telepathy (see Chapter 5), mediumship (see Chapter 2), and psychokinesis (see Chapter 9), suggesting that it may be important, in some cases, to develop the capacity to maintain a concentrated, single-pointed attention.

Mindfulness Meditation

While this practice has gained a lot of attention in recent years, the exact definition of mindfulness is still being debated. Following the example of Jon Kabat-Zinn, the founder of Mindfulness-Based Stress Reduction (MBSR), we define mindfulness as "paying attention in a particular way, on purpose, in the present moment, and without judgment" (2009). It is the ability to observe things (internally and externally) without creating stories about them. Mindfulness involves experiencing thoughts, feelings, bodily sensations, and sensory inputs as pure sensations: no interpretations, no judgments, no analysis. These practices tend to quiet down and relax attention in the frontal lobe regions, creating somewhat of a separation from the typical way the mind tends to operate. Because mindfulness practices teach nonattachment, these approaches are powerful tools when working to manage anxiety and stress.

From a psi perspective, mindfulness may be the perfect tool to engage an attentional process a bit differently than that explored in focus practices. Our attention during mindfulness is more relaxed and open. It is an allowing, unrestricted, flowing form of attention. This is likely to be a similar style of attention to that observed with mediums and psychics who showed a reduction of activity in the frontal lobes. Rather than orienting attention to a single object, which requires activation, mindfulness provides a broad attention to anything (and everything) that moves across the field of awareness. You might recall that Laura Lynne Jackson showed a significant decrease of activity in the right frontal lobe during both psychic and mediumship readings (see Chapter 2).

Open Heart Meditation

These practices involve activating a positive feeling state and then directing those feelings toward self or others. Practices such as loving-kindness, compassion, gratitude, and forgiveness fit in this category. These practices can be very helpful for anyone dealing with resentment, unresolved grief, anger management, or anyone simply wanting to be better at understanding others.

When you experience a "felt sense" of the positive emotions that are encouraged in these practices, you activate specific portions of the right hemisphere (insula). This part of the brain is important for empathy and perspective taking, which may be a key element to certain psychic tasks, such as telepathy, as well as PK and energy healing. The insula is located in the temporal lobe (just above the ear) and was an area that became engaged when Joanne Gerber was involved in a psychic reading. The practice of open-heart meditation may also facilitate changes in the right parietal lobe (RPL), which is also involved in perspective taking and boundaries between self and other. In fact, certain open-heart practices, such as tonglen, invite the meditator to take on the suffering of others, transmute that energy, and then send out feelings of compassion, love, and well-being.

The RPL was a consistently accessed brain region throughout this book, suggesting that the ability to intentionally soften ego boundaries is a key element to many psi abilities.

Quiet Mind Meditation

Practices in this category represent the stereotype of meditation. This is a state in which internal chatter has been reduced to a minimum. Sometimes it is described as a feeling of spaciousness or emptiness. This state is common in traditions like Zen or Transcendental Meditation (TM). Not surprisingly, the brain patterns connected to these practices show a significant quieting of many regions of the brain, including the default mode network (DMN) and language centers. Because these practices essentially involve interrupting the "normal" process of "selfing," these practices can be very helpful for any concerns connected to a distorted or inaccurate perception of self.

Nearly every practitioner I spoke to while writing this book emphasized the importance of quieting the mind. To be more precise, they seem to be describing a reduction of activity in the thinking, analytic mind; the mind that wants to have an internal conversation about everything that is happening. The fact that the DMN becomes less active during these practices shows us that the tendency to think about your identity or your "self" during these practices is reduced. This may provide an opportunity to tune in to information beyond the self.

Deep State Meditation

Meditations in this category involve letting go of attentional control while in a deep state of physical and mental relaxation. These practices can include dreamlike imagery, transcendent experiences, and activation of subconscious material. Practices that facilitate this state of consciousness assist in the development of openness, surrender, lowering of psychological defenses, and creativity. The connection between this type of meditative state and psi abilities should be obvious. This is a state in which the normal screening process of the brain is temporarily inhibited, allowing hidden or otherwise subconscious material to come to the surface. This state shares some similarities with psychedelic states, although not as dramatic or intense. We saw examples suggestive of a deep state meditation during some of the technology-based approaches. For example, when Julie Rost described her experience with rTMS as "psychedelic" (see Chapter 8), or in the Ganzfeld experiments when Cammra Garza described a vision-like experience (see Chapter 3).

Altered States of Consciousness

Experimenting with altered states of consciousness (ASC) may facilitate the development of psi abilities in at least two ways. First, there are stories like Janet Mayer's, who after an ASC experience with holotropic breathwork unlocked abilities that she did not have beforehand. In her case, she started spontaneously channeling voices of shamans from South America. How and why this kind of big shift happens in some cases is not clear but suggests that the ASC somehow opened or unlocked this ability. In fact, if you refer back to Chapter 2, we found that a specific region of Janet's brain would essentially go offline when she was engaged in channeling.

The other way that ASCs may increase psi abilities is by increasing the brain's flexibility or entropy. This increased flexibility serves a purpose similar to the deep states meditations by temporarily removing normal filters in consciousness and allowing more subtle information to come through. For most of us, we spend much of our waking consciousness leaning toward a low entropy state. This is probably useful while navigating the physical world as it allows us to block out distractions and stay focused on what is right in front of us. However, this also prevents us from tapping into subtle sources of information. While ASCs can be achieved through various means—including breathwork, sensory deprivation (float tanks), sensory overload (audio-visual entrainment), vibroacoustics, and others—most ASC/psi-related research comes from the field of psychedelic medicines, such as LSD and psilocybin.

In an article published in the *International Journal of Transpersonal Studies*, David Luke (2012) provides a comprehensive review of the literature related to psychedelics and psi phenomena, noting that there have been more than two hundred such publications. In the beginning of the article, he reports that psi-related experiences have been studied in relation to psychedelics since the 1950s. Luke goes on to list eighteen qualities of the psychedelic state that seem related to psi experiences. These include: transcendence of space and time, release of unconscious material into the conscious mind, increased openness, increased optimism towards impossible realities, intensity of feeling, and sensitivity of subtle changes (to name a few).

In addition to the immediate impact of ASCs on expanding consciousness, it is noted that psychedelics may also result in long-term changes that may be *psi conducive*. For example, it is not uncommon for someone to have a profound psychedelic experience that leads to a (more or less) permanent change in the person's ideas and beliefs about reality, leading to greater openness to unusual experiences. I have personally heard firsthand accounts from dozens of psychedelic users describing experiences that have convinced them that there is much more to our existence than can be explained by this physical body. Friends and colleagues have told me stories about being connected and feeling "One" with everything in the Universe, meeting angels and spirit guides, communicating with ancestors, engaging with aliens, astral travel, psychic surgery, transforming into plants or animals, being able to manipulate energy, and see into the future. Inevitably, when they relate these experiences, they describe them as feeling "more real" than our conversation.

Of course, as noted elsewhere in this book, exploring ASCs is not for everyone and should be approached thoughtfully and carefully. Regardless of the induction technique, ASC experiences can be intense and overwhelming if you are not properly prepared. It is strongly recommended that any ASC work be explored with an experienced teacher or guide in a monitored setting. In a perfect world this experience would include significant preparation time. Preparation sessions can be helpful in assessing your readiness for the experience: clarifying intentions, managing expectations, outlining the process, as well as exploring any fears or concerns you may have. After the ASC session, you should spend several weeks actively engaged in integration of the experience. These sessions can assist in processing and synthesizing the experience, and in identifying and enacting steps toward implementing any insights received during your altered state of consciousness. This is a time to take advantage of the temporary neural plasticity achieved through the ASC and clearly work toward your goals.

Technology-Based Interventions

Using brain-based biofeedback (neurofeedback) and/or neurostimulation technologies it is possible to train and/or entrain the brain in specific ways. Our preliminary research suggests that these approaches can be used strategically to help the brain become more flexible and open, potentially exaggerating our natural psi abilities.

I recommend using one or more of the approaches detailed below alongside consistent psi practice and mentoring with an experienced psychic, medium, or energy healer.

EEG-Guided Meditation/Neuromeditation

Beginning meditators often complain that they do not know if they are "doing it right," or give up before realizing its significant benefits. Advanced meditators often reach a plateau and struggle to reach the next or deeper level of their practice. Traditionally, meditation practitioners have sought a guru or *sangha* for support and direction. Yet many people are not interested in or have access to that kind of approach. A potential solution to these challenges involves using brain biofeedback to increase awareness of subtle states of consciousness and speed the meditation learning process.

Simply put, biofeedback is any kind of mechanism that allows you to receive information (feedback) about your physiology (bio). This can include heart rate, skin temperature, or other physiological measures of nervous system arousal. With brain-wave biofeedback, or EEG biofeedback, it is possible to monitor brain activity and use this information to gently guide the user into specific states of meditation. This process is generally called *neuromeditation*.

By tracking brain wave activity in specific regions of the brain, we can tell if someone is focused or relaxed. We can tell if the mind is wandering, if they are engaged in body-based emotions, or if they have entered a space of internal quiet. By monitoring this activity and connecting it directly to the intent of the meditation, it is possible to help meditators learn to quickly enter a desired state of consciousness and maintain this state for increasing periods of time.

Let me provide a simple example of how this works with a focus style of meditation. In this type of meditation, a person directs attention toward a single target, such as the breath, a candle flame, a mantra, or the third eye. When the mind inevitably wanders, they recognize that the mind wandered and gently return attention to the target. By monitoring brain-wave activity in specific regions of the brain, we can get a pretty good idea if the person is actually engaged in the meditation or is instead caught in mind wandering. When the attention is focused, they receive an auditory reward, such as an increase in volume of ambient meditation music. This feedback indicates to the person that they are on track. When the mind wanders and the brain waves shift out of the desired pattern, the volume of the music decreases, providing direct and nearly immediate feedback to the meditator, allowing them to refine their internal awareness.

Historically, the cost of the required equipment and the necessary training involved has made this approach inaccessible to nonneu-rofeedback practitioners. However, recent advances have changed all of that. Now, there are Bluetooth, wireless, dry sensor, and EEG hardware solutions that make the hookup very easy. For example, the BrainBit Flex EEG cap allows you to move the four sensors to any location, providing the flexibility necessary for many different neuromeditation approaches.

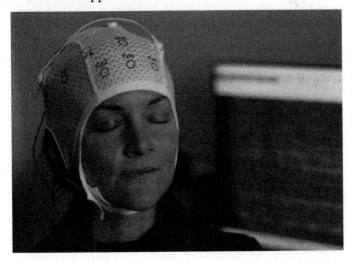

This type of device can be used with Divergence Neuro software, which allows you to access the neuromeditation programs right from your smart phone or tablet. You can find out more about where to find these products in the Resources section.

Audio-Visual Entrainment

Audio-Visual Entrainment (AVE) technologies were discussed in Chapters 3 and 6. This technique uses both sound and light stimulation to impact brain wave patterns (Siever and Collura, 2017). Most AVE devices consist of a set of eyeglasses with lights built into the eyesets and headphones. The glasses and headphones are connected to either a small controller box or to the computer. Built-in programs in the software allow you to choose from a variety of settings that will cause the lights in the glasses to flicker on and off at specific frequencies. The tones in the headset will follow the same pattern. Once you choose your program, you sit back, close your eyes, and enjoy the show for fifteen to twenty-five minutes.

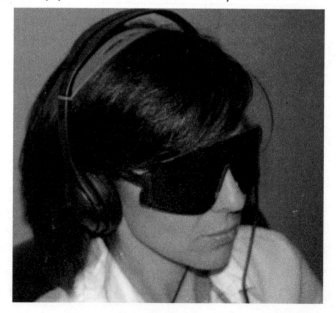

The idea is that by providing the brain with a consistent repetitive signal, it will follow along (entrainment). For example, if I chose a relaxation program, the lights might flicker on and off ten times a second. The brain responds to each of these signals, producing a ten-cycles-per-second brain wave (10 Hz), which is in the alpha frequency range associated with relaxed, internal attention. Using this technology, you can influence the brain by encouraging it toward certain brain-wave patterns, which can influence your state of consciousness.

These devices and their specific programs are generally used for relaxation, to help with sleep, to increase feelings of alertness, to improve mood, or as a treatment option for a wide variety of mental health concerns. For example, many people with anxiety or chronic stress often have an excessive amount of fast brain wave activity (Price and Budzynski, 2009; Olbrich et al., 2011), causing the mind to be overactive; that is, thinking and worrying more than is helpful. An AVE program to counterbalance this hyperactivity would encourage the brain toward slower brain waves (7–10 Hz), leading to feeling centered and relaxed.

People suffering from depression, ADHD, or cognitive decline often have the opposite challenge. Their brains may be producing an excessive amount of slow brain waves, causing them to feel tired, unmotivated, inattentive, impulsive, or mentally "foggy" (Fernández-Palleiro et al., 2020; Garcia et al., 2019; Jeong, 2004). AVE protocols designed to speed the brain up a bit (14–18 Hz) can often help pull the brain out of this pattern, resulting in a feeling of having more energy, optimism, executive function, and self-control.

AVE technologies are often used as an at-home adjunct to neurofeedback therapy. However, they can also be used to facilitate certain meditations or nonordinary states of consciousness. For example, some meditations often lead to an increase of brain waves between 8 and 10 Hz (alpha1). By using an AVE protocol that stimulates 9 Hz activity while you are practicing certain meditation approaches, you can provide a boost to that specific meditative state (Tarrant, 2017). Protocols that regularly shift frequencies and colors, or maintain a solid color for an extended time (Ganzfeld), can also be used to induce altered states of consciousness (Wackermann et al., 2008).

While there is solid research behind this technique, it is important to note that results may vary. Everyone is different and it is impossible to know what approaches may or may not help without an appropriate assessment and some trial and error. The good news is that people tend to notice any change in mood or mental state soon after a session is completed. By experimenting with various protocols, light intensity, time of day, and length of session, it is usually possible to find one or more approaches to help balance the nervous system and/or shift into desired meditative/ASC states of awareness.

Pulsed Electromagnetic Frequencies

Pulsed electromagnetic field technologies emit electromagnetic waves at different frequencies to affect the functioning of the body and brain. Put simply, your cells require a certain amount of electrical energy (voltage) to function properly. We can influence these energy levels through strategic exposure to specific electromagnetic (EMF) frequencies.

Various researchers have demonstrated that toxins, aging, and chronic stress diminish the voltage in cells, inhibiting their ability to do their jobs effectively and efficiently. In fact, people with chronic illnesses consistently show seriously compromised cellular power. As Bryant A. Meyers, author of the book, *PEMF: The Fifth Element of Health,* put it, "If your cells have energy, you are healthy, if not, you are sick." (PEMF stands for Pulsed Electromagnetic Field.) Exposure to specific frequencies can boost the health and functioning of these cells. The frequencies that seem most beneficial are those that are associated with the natural frequencies of the Earth.

The Earth's EMF Field(s)

Without getting too technical, the Earth generates its own electromagnetic field (EMF). According to the United States Geological Survey, "The Earth's outer core is in a state of turbulent convection as the result of radioactive heating and chemical differentiation. This sets up a process that is a bit like a naturally occurring electrical generator, where the convective kinetic energy is converted to electrical and magnetic energy" (https://www.usgs.gov). While the frequencies generated by the Earth vary, the dominant frequency is around 11.75 Hz. In addition to the Earth's EMFs, there is another source of EMFs that exists between the Earth and the ionosphere. This energy is generated primarily by lightning strikes, solar flares, and other energy discharges. This field is referred to as the Schumann Resonance and has a dominant frequency around 7.83 Hz.

The EMFs generated by the Earth and the Schumann Resonance create an EMF field that all living beings on this planet are bathed in. These are the frequencies of nature. When we are in tune with these frequencies, the cells of the brain and body are more energized and productive. Consequently, spending time in nature may be one of the best forms of medicine. However, as a supplement it can be helpful to increase exposure to these Earth-based frequencies through PEMF technology. The top eight benefits of this approach as reported by Bryant Meyers include: stronger bones, endorphin production, pain relief, better sleep and human growth hormone (HGH) secretion, more energy and adenosine triphosphate (ATP) production (ATP powers cellular metabolism), better oxygenation/

circulation, improved immunity, relaxation and stress reduction, and nerve/tissue regeneration.

PEMF and Consciousness

While the Earth's EMF fields have dominant frequencies (7.83 and 11.75), they actually produce a wide range of frequencies that generally range from 0 to 30 Hz—the same basic frequency range of the brain. As we have seen throughout this book, changing the behavior of frequencies in the brain can have a significant impact on consciousness. By using PEMF devices to intentionally direct specific frequencies to the brain, we can influence consciousness, slowing things down, speeding them up, or disrupting their normal functioning. This approach was featured and described in Chapter 6 (Zapping the Brain). In that chapter we discussed the work of Michael Persinger and saw that a consumer-grade PEMF device (Neo-Rhythm) could be used to stimulate and exaggerate psi-related brain activity, resulting in an enhanced experience. When used for these purposes, it has been my experience that it is best to use the device during meditation and then practice your psi abilities afterward. Of course, everyone is different. Like the approach recommended with AVE, it may be necessary to experiment with various programs, placements, length of sessions, and timing of practice to achieve optimal results.

Lessons Learned

The process of writing this book has shown me that humans are capable of much more than most of us have ever imagined. While we may not fully understand how or why, it seems clear to me now that psi abilities are a natural and normal part of existence. When these abilities are inconsistent, messy, vague, and complicated, it is because our brains are interfering. The brain filters, limits, and distorts the information we receive as a strategy to help us get by on this planet. In that regard, it is helpful. If we did not have filters, limits, and meaning-making cognitive distortions, we would be overwhelmed with sensory information. Perhaps the challenge is understanding how the mind works and learning how to intentionally navigate varied states of consciousness. Rather than being stuck in a narrow perceptual state, we can develop cognitive and psychological flexibility. We can learn to move smoothly between logical, structured, analytic processing, and the subtle, mysterious, intuitive ways of understanding. We can be scientists *and* mystics, skeptics *and* believers.

RESOURCES

Psychics, Mediums, and Healers (in order of appearance)

Janet Mayer

Janet is a founding research medium and volunteer with the Forever Family Foundation. She provides psychic medium readings to bring comfort to those grieving. With her validating information to clients, she helps them to see that although their loved ones have left the physical body, they are still present and continue to interact with them in many ways. You can listen to Janet on the fourth Thursday of the month on the *Signs of Life* radio show provided by Forever Family Foundation, where she shares her personal insight and knowledge with callers. Janet is the author of *Spirits . . . They Are Present.*
www.JanetMayer.net

Kim Russo

Kim is a certified medium with the Forever Family Foundation and The Windbridge Institute for Applied Research and Human Potential. Kim's evidential readings have led to her success hosting six seasons of *The Haunting Of*, the television series *Psychic Intervention,* and the A&E weekly series Celebrity Ghost Stories. Kim is the author of *The Happy*

Medium: Life Lessons from the Other Side and Your Soul Purpose: Learn How to Access the Light Within.
www.kimthehappymedium.com

Laura Lynne Jackson

Laura serves as a Windbridge-Certified Research Medium and is also certified with the Forever Family Foundation. Her work as an evidential medium has been featured in numerous books, television shows, and series, including *The Dr. Oz Show, The Goop Lab*, and the Netflix series, *Surviving Death*. She is the *New York Times* bestselling author of *The Light Between Us* and *Signs*.
www.lauralynnejackson.com

Joanne Gerber

Joanne is an integrative research medium with the University of Arizona VERITAS research program. She is also certified through Windbridge Research Center and the Forever Family Foundation. Joanne offers psychic and medium telephone sessions, classes, workshops, and events in person and online.
www.joannegerber.com

Angelina Diana

Angelina is certified by the Forever Family Foundation and was a host of the radio show *Signs of Life* from 2006–2013. She provides future reading and spiritual counseling. She offers evidence-based spirit connection readings for those in grief, with services in person or by video chat. She also offers lectures, classes, and group sessions.
www.angelinadiana.com

Rebecca Anne LoCicero

Rebecca Anne is a certified medium with the Forever Family Foundation providing evidential spirit communication through readings. She also provides psychic services as well as energy healings. Rebecca is a public speaker and expert in the field of metaphysics. She is the author of the book *Living with Messages from Heaven*.
www.RebeccaAnneLoCicero.com

Cammra Garza

Cammra is a psychic, medium, teacher, and cross-cultural shamanic practitioner in Eugene, Oregon. She works with private clients and small groups in person at Woodland Healing and remotely by phone or video chat.

www.woodlandhealing.com

Jeannine Kim

Jeannine is a mystic-intuitive, medium, astrologer, and energy healer. She is a dedicated student, teacher, and truth-seeker. Jeannine offers private readings, mentoring, group sessions, and Into the Mystic psychic development training. She is the author of the book *Dark Matters*.

www.jeanninekim.com

Julie Rost

Julie serves up not only a great cup of coffee, but also a big cup of information from departed loved ones, tarot cards, and a whole team of guides, angels and the rest of her Divine support team. Julie offers readings at Community Coffee in Eugene, Oregon, or over the phone.

www.communitycupcoffee.net

Ken Cohen aka Gao Han

Ken is a Qigong and Tai Chi Grandmaster, culture educator, and traditional healer. Master Ken offers lectures, workshops, classes, and teacher training as well as healing and consultations in Colorado, California, and throughout the world. He is the author of *The Way of Qigong* and *Honoring the Medicine*

www.qigonghealing.com

Katherine Glass

Katherine is the cofounder of the Healing Essence Center in Concord, Massachusetts. She is a spiritual counselor, psychic medium, and energy healer. Katherine is a graduate of the Barbara Brennan School of Healing and Sharon Turner's Awakenings Clairvoyant Program. She offers energy healing, psychic intuitive readings, and mediumistic readings.

www.katherineglass.com

Glenn Mendoza, MD

Master Glenn is a neonatologist, and Master Pranic Healer, teacher, and speaker. One of eight Master Pranic Healers in the world, Master Glenn travels the world teaching and offering his wisdom to help ease suffering and enhance people's spiritual connection through the teachings and workshops of Grandmaster Choa Kok Sui. He is the author of *Better Person Mindset, Spiritual Truths*, and *Rules that Guide Us*.
www.pranichealingusa.com

Robert Allen aka Trebor Seven

Robert is a self-taught practitioner and freelance teacher of psychokinesis. He offers education and training on telekinesis (moving objects), aerokinesis (moving air), atmokinesis (weather control), biokinesis (biological organisms), hydrokenisis (water), pyrokinesis (moving fire) and other psychic abilities.
www.TelekinesisSchool.com
www.TreborSeven.com
www.SacredSupplements.com

Caroline Cory

Caroline is an award-winning filmmaker, executive producer, and founder of Omnium Media, a media platform tackling thought-provoking topics on the human condition and the nature of reality. In addition to writing and producing, Caroline lectures and coaches internationally on various mind-over-matter subjects. Caroline is the film producer of *A Tear in the Sky* (2022), and *Superhuman: The Invisible Made Visible* (2020).
www.omniumuniverse.com

Researchers and Organizations

Psychic Mind Science (PMS)

Founded by Dr. Jeff Tarrant, PMS is devoted to exploring the interface of science and psi-related abilities. Research activities are focused on examining brain and biofield changes related to mediumship, telepathy,

psychokinesis, and energy healing. Online classes and in -person retreats teach meditation techniques and technology-based interventions to activate or inhibit brain areas involved in psi-activities.
www.PsychicMindScience.Com

NeuroMeditation Institute (NMI)

The workshops and classes taught by NMI are founded on the understanding that each style of meditation has a different impact on the structure and function of the brain. This knowledge can be used to identify which meditation practices are most likely to help you achieve mental wellness and enhance your ability to access your psychic potential. NMI offers both remote and live classes, professional training, and technology-based tools to improve mental health and wellness.
www.NeuroMeditationInstitute.Com

Marilyn Schlitz, PhD, MA

Dr. Schlitz is an acclaimed social scientist, award winning author, and charismatic public speaker. She has conducted clinical, laboratory, and field-based research into consciousness, human transformation, and healing. She is currently Professor of Transpersonal Psychology at Sofia University and Senior Fellow at the Institute of Noetic Sciences (IONS). Dr. Schlitz has published hundreds of articles in scholarly journals and is the author of *Living Deeply: The Art & Science of Transformation in Everyday Life, Consciousness and Healing: Integral Approaches to Mind Body Medicine,* and *Death Makes Life Possible.*
https://marilynschlitz.com/

Diane Hennacy Powell, MD

Dr. Powell is an integrative medical doctor, neuropsychiatrist, and psychotherapist. Dr. Powell completed her training at Johns Hopkins University School of Medicine and was on faculty at Harvard Medical School before leaving academia for private practice. Her research with telepathy and autistic savants is the subject of a forthcoming docuseries. In addition, she is the author of *The ESP Enigma: The Scientific Case for Psychic Phenomena.*
https://drdianehennacy.com/

Forever Family Foundation (FFF)

Forever Family Foundation is an all-volunteer, not-for-profit organization with 13,000 members in 76 countries. Its mission is to educate the public about evidence suggesting that we survive physical death and offer support to the bereaved.

www.foreverfamilyfoundation.org

Windbridge Research Center

The mission of the Windbridge Research Center is to ease suffering around dying, death, and what comes next by performing rigorous scientific research and sharing the results and other customized content with the general public, clinicians (like medical and mental health professionals), scientists (like researchers and philosophers), and practitioners (like mediums).

www.windbridge.org

Rhine Research Center

The Rhine Research Center explores the frontiers of consciousness and exceptional human experiences in the context of unusual and unexplained phenomena. The Rhine's mission is to advance the science of parapsychology, to provide education and resources for the public, and to foster a community for individuals with personal and professional interest in PSI.

www.rhineonline.org

Institute of Noetic Science (IONS)

The Institute of Noetic Sciences (IONS) is inspired by the power of science to explain phenomena not previously understood, harnessing the best of the rational mind to make advances that further our knowledge and enhance our human experience. The scientists at IONS apply the rigors of their respective disciplines to explore the mysteries of human potential with a focus on understanding humanity's inherent interconnectedness and the inner wisdom common to us all.

www.noetic.org

NOTES

Chapter 1: Channeling Shamans

Beauregard, M., & Paquette, V. (2006). "Neural Correlates of a Mystical Experience in Carmelite Nuns." *Neuroscience Letters*, *405*(3), 186–190.

Bonny, H. L., & Pahnke, W. N. (1972). "The Use of Music in Psychedelic (LSD) Psychotherapy." *Journal of Music Therapy*, *9*(2), 64–87.

Gerbarg, P. L., & Brown, R. P. (2005). "Yoga: A Breath of Relief for Hurricane Katrina Refugees." *Current Psychiatry*, *4*(10), 55–67.

Eyerman, J. (2014). "Holotropic Breathwork: Models of Mechanism of Action." *Journal of Transpersonal Research*, *6*(1), 64–72.

Kaelen, M., Giribaldi, B., Raine, J., et al. (2018). "The Hidden Therapist: Evidence for a Central Role of Music in Psychedelic Therapy." *Psychopharmacology*, *235*, 505–519.

Johnstone, B., Bodling, A., Cohen, D., Christ, S. E., et al. (2012). "Right Parietal Lobe-Related 'Selflessness' as the Neuropsychological Basis of Spiritual Transcendence." *International Journal for the Psychology of Religion*, *22*(4), 267–284.

Johnstone, B., & Glass, B. A. (2008). "Support for a Neuropsychological Model of Spirituality in Persons with Traumatic Brain Injury." *Zygon*, *43*(4), 861–874.

Kass, J. (2000). Index of Core Spiritual Experiences (INSPIRIT). Lesley University, Cambridge, MA.

Grof, S. (1980). *LSD Psychotherapy*. Hunter House Publishers, Nashville, Tennessee.

Rhinewine, J. P., & Williams, O. J. (2007). "Holotropic Breathwork: The Potential Role of a Prolonged, Voluntary Hyperventilation Procedure as an Adjunct to Psychotherapy." *The Journal of Alternative and Complementary Medicine, 13*(7), 771–776.

Uddin, L. Q., Molnar-Szakacs, I., Zaidel, E., & Iacoboni, M. (2006). rTMS to the right inferior parietal lobule disrupts self–other discrimination. *Social Cognitive and Affective Neuroscience, 1*(1), 65–71.

Urgesi, C., Aglioti, S. M., Skrap, M., et al. (2010). "The Spiritual Brain: Selective Cortical Lesions Modulate Human Self-Transcendence." *Neuron, 65*(3), 309–319.

Wise, A. (2002). *Awakening the Mind: A Guide to Mastering the Power of Your Brainwaves*. Jeremy P. Tarcher/Putnam, New York.

Chapter 2: The Mind of the Medium

Barnett, L., Muthukumaraswamy, S. D., Carhart-Harris, R. L., & Seth, A. K. (2020). "Decreased Directed Functional Connectivity in the Psychedelic State." *NeuroImage, 209*, 116462.

Lehmann, D., Faber, P. L., Tei, S., Pascual-Marqui, R. D., et al. (2012). "Reduced Functional Connectivity Between Cortical Sources in Five Meditation Traditions Detected with Lagged Coherence Using EEG Tomography." *NeuroImage, 60*(2), 1574–1586.

Smigielski, L., Scheidegger, M., Kometer, M., & Vollenweider, F. X. (2019). "Psilocybin-Assisted Mindfulness Training Modulates Self-Consciousness and Brain Default Mode Network Connectivity with Lasting Effects." *NeuroImage, 196*, 207–215.

University of California-Los Angeles (2013, October 17). "Psychologists Report New Insights on Human Brain, Consciousness." *ScienceDaily*. Retrieved from http://www.sciencedaily.com/releases/2013/10/131017173646.htm#.UmL05CFtd0U.gmail

Chapter 3: Ganzfeld

Radin, D. (2009). *Entangled Minds: Extrasensory Experiences in a Quantum Reality*. Simon & Schuster, New York.

Chapter 4: Gamma Waves and ESP

Acosta-Urquidi, J. "EEG Gamma Oscillations' Role in Healthy Brain Function, Neuropathology, and Altered States of Consciousness." In *Advances in Psychology Research*, Vol. 137, 2019. Columbus, A.M. (Ed.), Nova Science: New York.

Braboszcz, C., Cahn, B. R., Levy, J., Fernandez, M., & Delorme, A. (2017). "Increased Gamma Brainwave Amplitude Compared to Control in Three Different Meditation Traditions." *PloS one*, 12(1), e0170647.

Honorton, C., & Ferrari, D. C. (1989). "Future Telling: A Meta-Analysis of Forced-Choice Precognition Experiments, 1935–1987." *Journal of Parapsychology*, 53(4), 281–308.

Judith, A. (2012). *Wheels of Life: A User's Guide to the Chakra System*. Llewellyn Worldwide, Woodbury, Minnesota.

Jung, C. G. (2010). Synchronicity: An acausal connecting principle. (From Vol. 8. of the collected works of CG Jung) (New in Paper) (Vol. 36). Princeton University Press.

Lutz, A., Greischar, L. L., Rawlings, N. B., Ricard, M., et al. (2004). "Long-Term Meditators Self-Induce High-Amplitude Gamma Synchrony During Mental Practice." Proceedings of the National Academy of Sciences, 101(46), 16369–16373.

Oakes, T. R., Pizzagalli, D. A., Hendrick, A. M., Horras, et al. (2004). "Functional Coupling of Simultaneous Electrical and Metabolic Activity in the Human Brain." Human Brain Mapping, 21(4), 257–270.

Radin, D. (2009). *Entangled Minds: Extrasensory Experiences in a Quantum Reality*. Simon & Schuster, New York.

Steinkamp, F., Milton, J., & Morris, R. L. (1998). "A Meta-Analysis of Forced-Choice Experiments Comparing Clairvoyance and Precognition." *Journal of Parapsychology*, 62(3), 193.

Strassman, R. (2000). *DMT: The Spirit Molecule: A Doctor's Revolutionary Research into the Biology of Near-Death and Mystical Experiences*. Simon & Schuster, New York.

Stuckey, D. E., Lawson, R., & Luna, L. E. (2005). "EEG Gamma Coherence and Other Correlates of Subjective Reports During Ayahuasca Experiences." *Journal of Psychoactive Drugs*, 37(2), 163–178.

Voss, U., Holzmann, R., Tuin, I., & Hobson, A. J. (2009). "Lucid Dreaming: A State of Consciousness with Features of Both Waking and Non-lucid Dreaming." *Sleep*, 32(9), 1191–1200.

YouGov NY Psychics and Mediums. US Nat. Sample: October 27th–30th, 2017. It was an online survey.

Chapter 5: Telepathy and Autistic Savants

Badran, B. W., Austelle, C. W., Smith, N. R., Glusman, C. E., et al. (2017). "A Double-Blind Study Exploring the Use of Transcranial Direct Current Stimulation (tDCS) to Potentially Enhance Mindfulness Meditation (E-Meditation)." *Brain Stimulation: Basic, Translational, and Clinical Research in Neuromodulation, 10*(1), 152–154.

Collesso, T., Forrester, M., & Baruš, I. (2021). "The Effects of Meditation and Visualization on the Direct Mental Influence of Random Event Generators." *Journal of Scientific Exploration, 35*(2).

Duane, T. D., & Behrendt, T. (1965). "Extrasensory Electroencephalographic Induction Between Identical Twins." *Science, 150*(3694), 367–367.

Farb, N. A., Segal, Z. V., Mayberg, H., Bean, J., et al. (2007). "Attending to the Present: Mindfulness Meditation Reveals Distinct Neural Modes of Self-Reference." *Social Cognitive and Affective Neuroscience, 2*(4), 313–322.

Hebb, D. O. (1949). *The Organization of Behavior: A Neuropsychological Theory*. Science Editions.

TED Radio Hour. (Feb. 20, 2015). *How Can a Stroke Change Your Brain?* NPR.

Kabat-Zinn, J. (2016). Defining Mindfulness. *Online: https://www.mindful. org/jon-kabat-zinn-defining-mindfulness.*

Kittenis, M., Caryl, P., & Stevens, P. (2004, August). Distant Psychophysiological Interaction Effects Between Related and Unrelated Participants. In *Proceedings of the Parapsychological Association Convention* (pp. 67–76).

Penberthy, J. P., Hodge, A. S., Hook, J. N., Delorme, A., et al. (2020). "Meditators and Nonmeditators: A Descriptive Analysis Over Time with a Focus on Unusual and Extraordinary Experiences." *Journal of Yoga Physiotherapy, 8*(3), 555744.

Powell, D. H. (2009). *The ESP Enigma: The Scientific Case for Psychic Phenomena*. Bloomsbury Publishing USA.

Radin, D. (2018). *Real Magic: Ancient Wisdom, Modern Science, and a Guide to the Secret Power of the Universe*. Harmony.

Radin, D., Michel, L., Galdamez, K., Wendland, P., et al. (2012). "Consciousness and the Double-Slit Interference Pattern: Six Experiments." *Physics Essays, 25*(2), 157.

Seidlmeier, P., Eberth, J., Schwarz, M., Zimmerman, D., et al. (2012). "The Psychological Effects of Mindfulness: A Meta-Analysis." *Psychological Bulletin, 138*, 1139–1171.

Standish, L. J., Johnson, L. C., Kozak, L., & Richards, T. (2003). "Evidence of Correlated Functional Magnetic Resonance Imaging Signals Between Distant Human Brains." *Alternative Therapies in Health and Medicine,* 9(1), 128.

Standish, L. J., Kozak, L., Johnson, L. C., & Richards, T. (2004). "Electroencephalographic Evidence of Correlated Event-Related Signals Between the Brains of Spatially and Sensory Isolated Human Subjects." *The Journal of Alternative and Complementary Medicine,* 10(2), 307–314.

Targ, R., & Puthoff, H. (1974). Information transmission under conditions of sensory shielding. *Nature,* 251(5476), 602–607.

Tarrant, J. (2017). *Meditation Interventions to Rewire the Brain: Integrating Neuroscience Strategies for ADHD, Anxiety, Depression and PTSD.* PESI Publishing, Eau Claire, Wisconsin.

Valk, S. L., Bernhardt, B. C., Trautwein, F. M., Böckler, A., et al. (2017). "Structural Plasticity of the Social Brain: Differential Change after Socio-Affective and Cognitive Mental Training." *Science Advances,* 3(10), e1700489.

Young, S. (2016). *The Science of Enlightenment: How Meditation Works.* Sounds True, Louisville, Colorado.

Chapter 6: Zapping the Brain

Blanke, O., & Arzy, S. (2005). "The Out-of-Body Experience: Disturbed Self-Processing at the Temporo-Parietal Junction." *The Neuroscientist,* 11(1), 16–24.

Blanke, O., Ortigue, S., Landis, T., & Seeck, M. (2002). "Stimulating Illusory Own-Body Perceptions." *Nature,* 419(6904), 269–270.

St.-Pierre, L., & Persinger, M. A. (2006). "Experimental Facilitation of the Sensed Presence Is Predicted by the Specific Patterns of the Applied Magnetic Fields, not by Suggestibility: Re-analyses of 19 Experiments." *International Journal of Neuroscience.*

Powell, D. H. (2009). *The ESP Enigma: The Scientific Case for Psychic Phenomena.* Bloomsbury Publishing, USA.

Tedrus, G. M., Vargas, L. M., & Rodrigues, K. G. (2023). "Religious Experience and Clinical-EEG Aspects in Adult People with Epilepsy." *Clinical EEG and Neuroscience,* 54(2), 198–202.

Uddin, L. Q., Molnar-Szakacs, I., Zaidel, E., & Iacoboni, M. (2006). "rTMS to the Right Inferior Parietal Lobule Disrupts Self–Other Discrimination." *Social Cognitive and Affective Neuroscience,* 1(1), 65–71.

Chapter 7: Psychomanteum

Barbato, M., Blunden, C., Reid, K., Irwin, H., et al. (1999). "Parapsychological Phenomena Near the Time of Death." *Journal of Palliative Care, 15*(2), 30–37.

Blanke, O., Ortigue, S., Landis, T., & Seeck, M. (2002). "Stimulating Illusory Own-Body Perceptions." *Nature, 419*(6904), 269–270.

Caputo, G. B. (2010). "Strange-Face-in-the-Mirror Illusion." *Perception, 39*, 1007–1008.

Caputo, G. B., Lynn, S. J., & Houran, J. (2021). "Mirror- and Eye-Gazing: An Integrative Review of Induced Altered and Anomalous Experiences." *Imagination, Cognition and Personality, 40*(4), 418–457.

Davies, R. (2004). "New Understandings of Parental Grief: Literature Review." *Journal of Advanced Nursing, 46*(5), 506–513.

Hastings, A. (2002). Guest Editorial: The Resistance to Belief. *Journal of Near-Death Studies, 21*, 77–98.

Hastings, A., Hutton, M., Braud, W., Bennett, C. et al. (2002). "Psychomanteum Research: Experiences and Effects on Bereavement." *OMEGA—Journal of Death and Dying, 45*(3), 211–228.

Kjellgren, A., Lyden, F., & Norlander, T. (2008). "Sensory Isolation in Flotation Tanks: Altered States of Consciousness and Effects on Well-Being." *The Qualitative Report, 13*(4), 636–656.

Klass, D., Silverman, P. R., & Nickman, S. (2014). *Continuing Bonds: New Understandings of Grief.* Taylor & Francis, UK.

Moody Jr., R. A. (1992). "Family Reunions: Visionary Encounters with the Departed in a Modern-Day Psychomanteum." *Journal of Near-Death Studies, 11*(2), 83–121.

Moody, R., & Perry, P. (1994). *Reunions: Visionary Encounters with Departed Loved Ones.* Ivy Books, Toronto, Ontario, Canada.

Neimeyer, R. A. (2014). "The Changing Face of Grief: Contemporary Directions in Theory, Research, and Practice." *Progress in Palliative Care, 22*(3), 125–130.

Parra, A., & Villanueva, J. (2010). "Unusual Perceptual Experiences and ESP under Psychomanteum Stimulation: Imagery/Hallucination Proneness and Schizotypal Personality Measures." *Australian Journal of Parapsychology, 10*(2), 120–138.

Radin, D. I., & Rebman, J. M. (1996). Are phantasms fact or fantasy? A preliminary investigation of apparitions evoked in the laboratory. *Journal-Society for Psychical Research, 61,* 65–87.

Schlitz, M. (2021, September 8). "From Grief to Growth: The Psychomanteum as a Tool for Exploring Expanded States of Consciousness." https://noetic.org/blog/grief-to-growth-psychomanteum/

Simmonds-Moore, C. (2021). Panel: After-Death Contact and Communication. SSE-PA Connections Conference. Online.

Terhune, D. B., & Smith, M. D. (2006). "The Induction of Anomalous Experiences in a Mirror-Gazing Facility: Suggestion, Cognitive Perceptual Personality Traits and Phenomenological State Effects." *The Journal of Nervous and Mental Disease, 194*(6), 415–421.

Uddin, L. Q., Molnar-Szakacs, I., Zaidel, E., & Iacoboni, M. (2006). "rTMS to the Right Inferior Parietal Lobule Disrupts Self-Other Discrimination." *Social Cognitive and Affective Neuroscience, 1*(1), 65–71.

Wassie, N. (2022). "Meditation-Induced After-Death Communication: A Contemporary Modality for Grief Therapy."

Chapter 8: Psychics and the Psychedelic Brain

Jerabek, I. (2022, Oct. 22) Does Meditation Increase Openness to, and Experiences of, the Paranormal? A New Study Investigates. *PRWeb.* https://www.prweb.com/releases/does_meditation_increase_openness_to_and_experiences_of_the_paranormal_a_new_study_investigates/prweb18975456.htm?fbclid=IwAR0DWAiMeDPAdXw2XPSny5TJ51CaT-738cuQhQE9U1FT4XCpiv7mVxSsRzoA

Berkovich-Ohana, A., Harel, M., Hahamy, A., Arieli, A., et al. (2016). "Data for Default Network Reduced Functional Connectivity in Meditators, Negatively Correlated with Meditation Expertise." *Data in Brief,* 8, 910-914.

Carhart-Harris, R. L., & Friston, K. J. (2019). REBUS and the anarchic brain: toward a unified model of the brain action of psychedelics. *Pharmacological reviews,* 71(3), 316–344.

Carhart-Harris, R. L., Leech, R., Hellyer, P. J., Shanahan, M., Feilding, A., Tagliazucchi, E., & Nutt, D. (2014). The entropic brain: a theory of conscious states informed by neuroimaging research with psychedelic drugs. *Frontiers in human neuroscience,* 20.

Corazza, O. (2010). "Exploring Space Consciousness & Other Dissociative Experiences: A Japanese Perspective." *Journal of Consciousness Studies,* 17(7-8), 173–190.

Kjellgren, A., Lyden, F., & Norlander, T. (2008). "Sensory Isolation in Flotation Tanks: Altered States of Consciousness and Effects on Well-Being." *The Qualitative Report,* 13(4), 636–656.

Lilly, J. C. (2007). *The deep self: Consciousness exploration in the isolation tank.* Gateways Books & Tapes: Nevada City, CA.

Luke, D. (2019). *Otherworlds: Psychedelics and Exceptional Human Experience.* Aeon Books, New York.

Luke, D. (2009, September). "Telepathine (Ayahuasca) and Psychic Ability: Field Research in South America." In Abstracts of the British Psychological Society, Transpersonal Psychology Section: 13th Annual Conference.

Luke, D. P., & Kittenis, M. (2005). "A Preliminary Survey of Paranormal Experiences with Psychoactive Drugs." *Journal of Parapsychology,* 69(2), 305.

Nour, M. M., Evans, L., Nutt, D., & Carhart-Harris, R. L. (2016). Ego-dissolution and psychedelics: validation of the ego-dissolution inventory (EDI). *Frontiers in human neuroscience,* 269.

Roseman, L., Nutt, D. J., & Carhart-Harris, R. L. (2018). Quality of acute psychedelic experience predicts therapeutic efficacy of psilocybin for treatment-resistant depression. *Frontiers in pharmacology,* 8, 974.

Speth, J., Speth, C., Kaelen, M., Schloerscheidt, A. M., Feilding, A., Nutt, D. J., & Carhart-Harris, R. L. (2016). Decreased mental time travel to the past correlates with default-mode network disintegration under lysergic acid diethylamide. *Journal of Psychopharmacology,* 30(4), 344–353.

Strassman, R., Wojtowicz, S., Luna, L. E., & Frecska, E. (2008). *Inner Paths to Outer Space: Journeys to Alien Worlds Through Psychedelics and Other Spiritual Technologies.* Simon & Schuster, New York.

Tart, C. T. (1993). "Marijuana Intoxication, Psi, and Spiritual Experiences." *Journal of the American Society for Psychical Research,* 87(2), 149–170.

Vivot, R. M., Pallavicini, C., Zamberlan, F., Vigo, D. et al. (2020). "Meditation Increases the Entropy of Brain Oscillatory Activity." *Neuroscience,* 431, 40–51.

Wilson, R. A. (1992). Cosmic Trigger, vol. 1. *New Falcon,* San Fransisco.

Chapter 9: Psychokinesis

Bösch, H., Steinkamp, F., & Boller, E. (2006). "Examining Psychokinesis: The Interaction of Human Iintention with Random Number Ggenerators—a Meta-Analysis." *Psychological Bulletin,* 132(4), 497.

Radin, D. (1985). "Pseudorandom Number Generators in Psi Research." *The Journal of Parapsychology,* 49(4), 303.

Radin, D. (2009). *Entangled Minds: Extrasensory Experiences in a Quantum Reality.* Simon & Schuster, New York.

Radin, D. (2018). *Real Magic: Ancient Wisdom, Modern Science, and a Guide to the Secret Power of the Universe.* Harmony.

Radin, D. I., & Nelson, R. D. (1989). "Evidence for Consciousness-Related Anomalies in Random Physical Systems." *Foundations of Physics,* 19(12), 1499–1514.

Radin, D. I., & Nelson, R. D. (2003). Meta-analysis of mind-matter interaction experiments: 1959-2000. *Healing, Intention and Energy Medicine.* London: Harcourt Health Sciences, 39–48.Williams, B. J. (2021). "Minding the Matter of Psychokinesis: A Review of Proof-and Process-Oriented Experimental Findings Related to Mental Influence on Random Number Generators." *Journal of Scientific Exploration,* 35(4), 829–932.

Chapter 10: Energy Healing

Astin, J. A., Harkness, E., & Ernst, E. (2000). "The Efficacy of 'Distant Healing': A Systematic Review of Randomized Trials." *Annals of Internal Medicine, 132*(11), 903–910.

Byrd, R. C. (1988). Positive therapeutic effects of intercessory prayer in a coronary care unit population.

Cohen, K. S. (1999). *"The Way of Qigong: The Art and Science of Chinese Energy Healing."* Wellspring/Ballantine, Toronto, Ontario, Canada.

Green, E. (1991). Copper Wall Research; Psychology and Psychophysics. *Subtle Energies & Energy Medicine Journal Archives, 10*(1).

Kelley, N. J., Hortensius, R., Schutter, D. J., & Harmon-Jones, E. (2017). "The Relationship of Approach/Avoidance Motivation and Asymmetric Frontal Cortical Activity: A Review of Studies Manipulating Frontal Asymmetry." *International Journal of Psychophysiology, 119,* 19–30.

Lutz, A., Greischar, L. L., Rawlings, N. B., Ricard, M. et al. (2004).

"Long-Term Meditators Self-Induce High-Amplitude Gamma Synchrony During Mental Practice." *Proceedings of the National Academy of Sciences, 101*(46), 16369–16373.

McTaggart, L. (2017). "*The Power of Eight: Harnessing the Miraculous Energies of a Small Group to Heal Others, Your Life, and the World.*" Simon & Schuster.

Roe, C. A., Sonnex, C., & Roxburgh, E. C. (2015). "Two Meta-Analyses of Noncontact Healing Studies." *Explore, 11*(1), 11–23.

Sicher, F., Targ, E., Moore 2nd, D., & Smith, H. S. (1998). "A Randomized Double-Blind Study of the Effect of Distant Healing in a Population with Advanced AIDS. Report of a Small-Scale Study." *Western Journal of Medicine, 169*(6), 356.

Sui, C. K. (2004). "*Miracles Through Pranic Healing: Practical Manual on Energy Healing.*" Energetic Solutions, Inc.

Chapter 11: Lessons Learned

Cohn, S. A. (1996). "Scottish Tradition of Second Sight and Other Psychic Experiences in Families."

Fernández-Palleiro, P., Rivera-Baltanás, T., Rodrigues-Amorim, D., Fernández-Gil, S. et al. (2020). "Brainwaves Oscillations as a Potential Biomarker for Major Depression Disorder Risk." *Clinical EEG and Neuroscience, 51*(1), 3–9.

Garcia, A., Gonzalez, J. M., & Palomino, A. (2019). "Identification of Patterns in Children with ADHD Based on Brain Waves." In *Human-Computer Interaction: 5th Iberoamerican Workshop, HCI-Collab 2019, Puebla, Mexico, June 19–21, 2019, Revised Selected Papers 5* (255–268). Springer International Publishing.

Garrison, K. A., Scheinost, D., Worhunsky, P. D., Elwafi, H. M. et al. (2013). "Real-Time fMRI Links Subjective Experience with Brain Activity During Focused Attention." *NeuroImage, 81*, 110–118.

Jeong, J. (2004). "EEG Dynamics in Patients with Alzheimer's Disease." *Clinical Neurophysiology, 115*(7), 1490–1505.

Kabat-Zinn, J. (2009). *Wherever You Go, There You Are: Mindfulness Meditation in Everyday Life*. Hachette, UK.

Luke, D. (2012). Psychoactive substances and paranormal phenomena: A comprehensive review. *International Journal of Transpersonal Studies, 31*(1), 12.

Lutz, A., Brefczynski-Lewis, J., Johnstone, T., & Davidson, R. J. (2008). "Regulation of the Neural Circuitry of Emotion by Compassion Meditation: Effects of Meditative Expertise." *PloS one, 3*(3), e1897.

Meyers, B. A. (2013). *PEMF—the Fifth Element of Health: Learn Why Pulsed Electromagnetic Field (PEMF) Therapy Supercharges Your Health Like Nothing Else!* Balboa Press.

Olbrich, S., Sander, C., Matschinger, H., Mergl, R., Trenner, M., Schönknecht, P., & Hegerl, U. (2011). "Brain and Body." *Journal of Psychophysiology.*

Piaget, J., & Cook, M. T. (1954). The development of object concept.

Plomin, R., & Deary, I. J. (2015). "Genetics and Intelligence Differences: Five Special Findings." *Molecular Psychiatry, 20*(1), 98–108.

Price, J., & Budzynski, T. (2009). "Anxiety, EEG Patterns, and Neurofeedback." *Introduction to Quantitative EEG and Neurofeedback: Advanced Theory and Applications*, 453-470. Insert: Siever & Collura, 2017, from page 284.

Siever, D., & Collura, T. (2017). Audio–visual entrainment: physiological mechanisms and clinical outcomes. In *Rhythmic stimulation procedures in neuromodulation* (pp. 51–95). Academic Press.

Tarrant, J. M. (2017). "NeuroMeditation: An Introduction and Overview." In *Handbook of Clinical QEEG and Neurotherapy* (96-113). Routledge, UK.

How Does the Earth's Core Generate a Magnetic Field? *USGS*, U.S. Department of the Interior.https://www.usgs.gov/faqs/how-does-earths-core-generate-magnetic-field#:~:text=The%20Earth's%20outer%20core%20is,to%20electrical%20and%20magnetic%20energy.

Wackermann, J., Pütz, P., & Allefeld, C. (2008). "Ganzfeld-Induced Hallucinatory Experience, Its Phenomenology and Cerebral Electrophysiology." *Cortex, 44*(10), 1364–1378.

Wahbeh, H., Radin, D., Yount, G., of Menie, M. A. W. et al. (2022). "Genetics of Psychic Ability—A Pilot Case-Control Exome Sequencing Study." *Explore, 18*(3), 264–271.

Zwir, I., Arnedo, J., Del-Val, C., Pulkki-Råback, L., Konte, B. et al. (2020). "Uncovering the Complex Genetics of Human Character." *Molecular Psychiatry, 25*(10), 2295–2312.

ABOUT THE AUTHOR

JEFF TARRANT, PHD, BCN, is the founder and director of Psychic Mind Science and the NeuroMeditation Institute in Eugene, Oregon. He is a licensed psychologist and board certified in neurofeedback with advanced training in psychedelic integration therapy.

Dr. Tarrant specializes in teaching, clinical applications, and research, combining technology-based interventions with meditative states for improved mental health and the expansion of psi-related abilities. His clinical work utilizes ketamine-assisted therapy, stroboscopic light, and neurofeedback to enhance psychological flexibility and overall well-being.

Dr. Tarrant's research examining the brainwave patterns of psychics, mediums, and energy healers has resulted in a unique understanding of how to develop our innate psychic capacities using meditation and neuromodulation technologies. Dr. Tarrant offers online classes, retreats, and individual training for psychic exploration and personal development.

In addition to his clinical background, Dr. Tarrant has also trained extensively in a variety of spiritual, meditative, and healing practices, including Zen, Vipassana, Taoism/Qigong, Arhatic Yoga/Pranic Healing, and Mindfulness-Based Stress Reduction (MBSR). His first book, *Meditation Interventions to Rewire the Brain*, provides a blueprint to choosing a meditation practice that addresses the specific needs of each individual.

In his spare time, Dr. Tarrant enjoys spending time with family, playing the drums, attending concerts, and hanging out in nature.

To learn more about Dr. Tarrant's research and upcoming courses, visit:

www.PsychicMindScience.Com

www.NeuroMeditationInstitute.Com

Follow on social media:

Instagram, Facebook, YouTube: @psychicmindscience